Knot Theory

The Carus Mathematical Monographs

Volume Twenty-four

KNOT THEORY

CHARLES LIVINGSTON

Indiana University–Bloomington

Published and Distributed by
THE MATHEMATICAL ASSOCIATION OF AMERICA

Mathematical Association of America
1529 Eighteenth Street, NW
Washington, DC 20036
800-331-1MAA   FAX: 202-265-2384

Complete Set ISBN 0-88385-000-1
Vol. 24 ISBN 0-88385-027-3

Printed in the United States of America

Current Printing (last digit):
10   9   8   7   6   5   4   3   2   1

THE
CARUS MATHEMATICAL MONOGRAPHS

Published by
THE MATHEMATICAL ASSOCIATION OF AMERICA

The following Monographs have been published:

To my parents,
Herbert and Rosetta Livingston

ACKNOWLEDGEMENTS

I thank Don Albers for suggesting that I write this book and for the remarkable patience he has since shown. Thanks are also due to my colleagues Jim Davis and Allan Edmonds for providing me with many valuable ideas and to Pat Gilmer and Richard Skora for their worthwhile suggestions. Comments from Thomas Banchoff, Marjorie Senechal, and Philip Straffin were also of great benefit. I am grateful for the careful readings by Swatee Naik, Joel Lash, Steve Paris, and Kevin Pilgrim. The advice of my brother Eric Livingston and my friend Elena Fraboschi have been invaluable, and I am greatly indebted to them for their constant support. Elena also assisted me with the many aspects of preparing the manuscript, and Mary Jane Wilcox helped me with much of the typing.

PREFACE

Knot theory is an unusual field. On the one hand, its subject matter is familiar to everyone; the most difficult questions concerning knots are easy to state and arise as naturally as any problems in mathematics. On the other hand, the subject seems quite different from those that usually fall into the realm of mathematics; even for trained mathematicians, it is often not clear how rigorous mathematical methods can be used to model the most basic questions concerning knots. This book describes some of the mathematical techniques of knot theory, and illustrates their application to a variety of problems.

The early chapters discuss how knotting can be given a formal mathematical description, present three of the basic methods of the theory, and then investigate the relationships among the methods. The exposition then moves to a study of properties of knots, including a detailed look at symmetries. Higher dimensional knotting is treated next. The book concludes with a survey of recent progress in combinatorial knot theory.

Mathematical prerequisites have been kept to a minimum. Basic linear algebra is used frequently and a familiarity with elementary group theory is called for occasionally. The exercises are an essential part of the exposition; many central ideas are developed there. More important, the exercises provide an opportunity to enjoy the experience of working in knot theory.

The goal is to present a cross-section of the many fascinating aspects of knot theory; topics have been chosen to demonstrate a diversity of techniques and their interplay, not to provide a complete survey. Proofs are used to illustrate the methods of the subject, and distracting technical arguments are usually summarized. The proofs that are presented range from detailed arguments to brief sketches. A survey of the many accessible sources on the subject is included in the references for those wanting to pursue the material in greater detail or breadth.

CONTENTS SUMMARY

There are three main parts to this book. The first part, comprising Chapters 2 through 5, develops the fundamentals of knot theory. Chapter 1 includes a discussion of the recent history of the study of knots. In the process some of the most interesting problems of knot theory are described. Chapter 2 focuses on the basic material of the subject, the precise definitions of knots and their deformations. It is here that one begins to see how mathematical methods can be applied to the study of knotting. The three main techniques of knot theory appear in the next chapters: Chapter 3 is devoted to combinatorial methods, Chapter 4 presents geometric techniques, and Chapter 5 illustrates algebraic tools. These chapters demonstrate the nature of the techniques and the types of problems to which each apply.

The second part presents advanced topics in knot theory. Chapter 6 describes relationships among the methods of the earlier chapters. The sources of these relationships are quite deep and subtle. As a consequence the work is delicate, but the results provide many new insights. In Chapter 7 several properties of knots are presented. The intention here is to describe some of the very natural questions that occur about knots and to illustrate how the methods developed so far can give detailed answers to these questions. Chapter 8 is devoted to the study of symmetry, one of the most beautiful properties of knotting. It is here that the tremendous power of the techniques developed earlier becomes most evident.

The third part is independent of the material of the second part. Two modern aspects of the subject are explored in these closing chapters. Chapter 9 provides an introduction to high dimensional knot theory and briefly indicates how the methods of classical knot theory can be applied. Chapter 10 describes new combinatorial methods. These methods greatly extend those of Chapter 3; the study of these combinatorial invariants is one of the most active and fascinating areas of knot theory today.

CONTENTS

CHAPTER 1:
A CENTURY OF KNOT THEORY

In 1877 P. G. Tait published the first in a series of papers addressing the enumeration of knots. Lord Kelvin's theory of the atom stated that chemical properties of elements were related to knotting that occurs between atoms, implying that insights into chemistry would be gained with an understanding of knots. This motivated Tait to begin to assemble a list of all knots that could be drawn with a small number of crossings. Initially the project focused on knots of 5 or 6 crossings, but by 1900 his work, along with that of C. N. Little, had almost completed the enumeration of 10-crossing knots. The diagrams in Appendix 1 indicate the kind of enumeration he was seeking.

Tait viewed two knots as equivalent, or of the same type, if one could be deformed to appear as the other, and sought an enumeration that included each knot type only once. The difficulty of this task is illustrated by the four knots in Figure 1.1. For now a knot can be thought of simply as a loop of rope. With some effort it is possible to deform the second knot to appear untangled, like the first. On the other hand, no amount of effort seems sufficient to unknot the third or fourth. Is it possible that with some clever manipulation the third could be transformed to look like the fourth? If a list of knots is going to avoid knots of the same type appearing repeatedly, means of addressing such questions are needed.

When Tait began his work in the subject, the formal mathematics needed to address the study was unavailable. The arguments that his lists were complete are convincing, but the evidence that the listed knots are distinct was empirical. Developing means of proving that knots are distinct remains the most significant of the many problems introduced by Tait.

Figure 1.1

Work at the turn of the century placed the subject of topology on firm mathematical ground, and it became possible to define the objects of knot theory precisely, and to prove theorems about them. In particular, algebraic methods were introduced into the subject, and these provided the means to prove that knots were actually distinct. The greatest success in this early period was the proof by M. Dehn in 1914 that the two simplest looking knots, the right- and left-handed trefoils, illustrated in Figure 1.2, represent distinct knot types; that is, there is no way to deform one to look like the other.

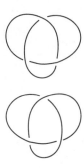

Figure 1.2

In 1928 J. Alexander described a method of associating to each knot a polynomial, now called the Alexander polynomial, such that if one knot can be deformed into another, both will have the same associated polynomial. This invariant immediately proved to be an especially powerful tool in the subject; a scan of Appendix 2 reveals that only 8 knots out of the 87 with 9 or fewer crossings share polynomials with others on the list.

Alexander's initial definitions and arguments were combinatorial, depending only on a study of the diagram of a knot, without reference to the algebra that had already proved so successful.

By 1932 the subject of knot theory was fairly well developed, and in that year K. Reidemeister published the first book about knots, *Knotentheorie*. The tools that he presented in the text are, in theory, sufficient to distinguish almost any pair of distinct knots, although as a practical matter for knots with complicated diagrams the calculations are often too lengthy to be of use.

One theme that was well established by this time was the study of families of knots. The most interesting family is formed by the torus knots, so called because they can be drawn to lie on the surface of a torus.

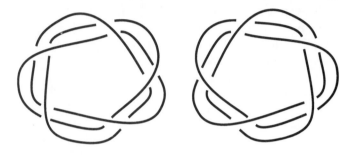

Figure 1.3

For any ordered pair of relatively prime integers, (p,q), with $p > 1$ and $|q| > 1$, there is a corresponding (p,q)-torus knot. Figure 1.3 illustrates the $(3,5)$-torus knot and the $(3,-5)$-torus knot. The right- and left-handed trefoils are easily seen to be the same as the $(2,3)$ and $(2,-3)$-torus knots, respectively. These knots provide test cases for new techniques and building blocks for constructing more complicated examples. Dehn and O. Schreier used group theoretic methods to give the first proof that the (p,q) and (p',q')-torus knots are the same if and only if the (unordered) sets $\{p,q\}$ and $\{p',q'\}$ are the same. (The Alexander polynomial of the (p,q)-torus knot turns out to be $(t^{|pq|} - 1)(t - 1)/(t^{|p|} - 1)(t^{|q|} - 1)$, and except for an issue of sign, this too is sufficient to distinguish the torus knots.)

Soon after *Knotentheorie* appeared, H. Seifert made a significant discovery. He demonstrated that if a knot is the boundary of a surface in 3-space, then that surface can be used to study the knot; he also presented an algorithm to

construct a surface bounded by any given knot. Figure 1.4 illustrates a surface with knotted boundary. This approach was certainly of practical importance, as it gave efficient means for computing many of the known invariants. More important, it laid the foundation for the use of geometric methods into a subject that, until

Figure 1.4

then, had been dominated by combinatorics and algebra.

In 1947 H. Schubert used geometric methods to prove a key result concerning the decomposition of knots. Given any two knots, one can form their connected sum, denoted $K \# J$, as illustrated in Figure 1.5. (If knots are thought of as being tied in a piece of string, the connected sum of two knots is formed by tieing them in separate por-

Figure 1.5

tions of the string so that they do not overlap.) A knot is called prime if it cannot be decomposed as a connected sum of nontrivial knots. (The appendix illustrates those prime knots with 9 crossings or less.) Schubert proved that any knot can be decomposed uniquely as the connected sum of prime knots. As an immediate corollary, if K is nontrivial, there is no knot J so that $J \# K$ is unknotted.

Unlike the problem of distinguishing knots, the problem of developing general means for proving that one knot *can* be deformed into another remained untouched. That changed in 1957. Early in the century Dehn gave an incorrect proof of what has become known as the Dehn Lemma. In rough terms, it stated that if a knot were indistinguishable from the trivial knot using algebraic methods, then the knot was in fact trivial. In 1957, C. Papakyriakopoulos succeeded in proving the Dehn Lemma, and it soon became the centerpiece of a series of major developments in the subject. One of special note occurred in 1968, when F. Waldhausen proved that two knots are equivalent if and only if certain algebraic data associated to the knots are the same. The interplay between algebra and geometry was essential to this work, and the connection was provided by Dehn's lemma.

The late 1950's through the 1970's were also marked by an extensive study of the classical knot invariants, and, in particular, how properties of the knot were reflected in the invariants. For instance, K. Murasugi proved that if a knot can be drawn so that the crossings alternate from over to under, then the coefficients of the Alexander polynomial alternate in sign. Figure 1.6 illustrates a non-alternating knot diagram—see how two successive over-

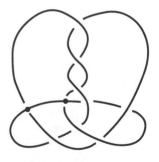

Figure 1.6

crossings are marked. By the Murasugi theorem, it is impossible to find an alternating diagram for this knot, as it has Alexander polynomial $2t^6 - 3t^5 + t^4 + t^3 + t^2 - 3t + 2$. Murasugi's work also detailed relationships between knot invariants and symmetries of knots, another major topic in the subject. Figure 1.7 illustrates three 9-crossing knots (9_4, 9_{17}, and 9_{33} in the appendix.) Two of the diagrams appear quite symmetrical, while the last is striking in its asymmetry. Is it possible to deform the third knot so that it too displays a similar symmetry?

Figure 1.7

In a completely different direction, the investigation of higher dimensional knots, such as knotted 2-spheres in 4-space, became a significant topic. In 1960 the subject consisted of little more than a sparse collection of examples. By 1970 it had become a well-developed area of topology. It also had become a significant source of questions concerning classical knots.

Since 1970, knot theory has progressed at a tremendous rate. J. H. Conway introduced new combinatorial methods which, when combined with more recent work by V. Jones, have led to vast new families of invariants. New geometric methods have been introduced by W. Thurston (hyperbolic geometry) and by W. Meeks and S. T. Yau (minimal surfaces), and together these have provided significant new insights and results. Finally, in 1988 C. McA. Gordon and J. Luecke solved one of the fundamental problems in knot theory. Many of the methods of knot theory focus not on the knot itself, but on the complement of the knot in 3-space; Gordon and Luecke proved that knots with equivalent complements are themselves equivalent.

Knot theory remains a lively topic today. Many of the basic questions, some dating to Tait's first paper in the subject, remain open. At the other extreme, the results of recent years promise to provide many new insights.

Figure 1.8

EXERCISES

1. If at a crossing point in a knot diagram the crossing is changed so that the section

that appeared to go over the other instead passes under, an apparently new knot is created. Demonstrate that if the marked crossing in Figure 1.8 is changed, the resulting knot is trivial. What is the effect of changing some other crossing instead?

Figure 1.9

2. Figure 1.9 illustrates a knot in the family of 3-stranded *pretzel knots*; this particular example is the $(5,-3,7)$ pretzel knot. Can you show that the (p,q,r)-pretzel knot is equivalent to both the (q,r,p)-pretzel knot and the (p,r,q)-pretzel knot?

3. The subject of knot theory has grown to encompass the study of links, formed as the union of disjoint knots. Figure 1.10 illustrates what is called the *Whitehead link*. Find a deformation of the Whitehead link that interchanges the two components. (It will be proved later that no deformation can separate the two components.)

Figure 1.10

4. For what values of (p,q,r) will the corresponding pretzel knot actually be a knot, and when will it be a link? For instance, if $p = q = r = 2$, then the resulting diagram describes a simple link of three components, "chained" together.

5.   Describe the general procedure for drawing the (p,q)-torus knot. What happens if p and q are not relatively prime?

6.   The link in Figure 1.11 is called the *Borromean link*. It can be proved that no deformation will separate the components. Note, however, that if one of the two components is removed, the remaining two can be split apart. Such a link is called *Brunnian*. Can you find an example of a Brunnian link with more than 3 components?

Figure 1.11

(H. Brunn described interesting families of such examples in 1892.)

7.   The knots illustrated in Figure 1.12 were, until recently, assumed to be distinct, and both appeared in many knot tables. However, Perko discovered a deformation that turns one into the other. Can you find it?

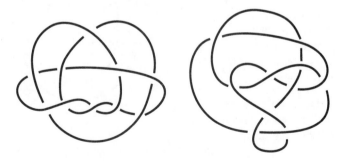

Figure 1.12

CHAPTER 2:
WHAT IS A KNOT?

There are many definitions of knot, all of which capture the intuitive notion of a knotted loop of rope. For each definition there is a corresponding definition of deformation, or equivalence. This chapter will concentrate on one pair of such definitions, and mention another. (Results at the foundations of geometric topology relate the various definitions. Such matters will not be presented here, and do not affect the work that follows.) The goal for now is to demonstrate how the notion of knotting can be given a rigorous mathematical formulation, and to give the reader a flavor of the problems and techniques that occur at this basic level of the subject.

1 Wild Knots and Unknottings

Considering a pair of definitions that are not appropriate, and seeing how they fail, demonstrates some unexpected subtleties and the need for precision and care in finding the right approach. One might define a knot as a continuous simple closed curve in Euclidean 3-space, R^3. To be precise, such a curve consists of a continuous function f from the closed interval $[0,1]$ to

R^3 with $f(0) = f(1)$, and with $f(x) = f(y)$ implying one of the three possibilities:

(1) $x = y$,

(2) $x = 0$ and $y = 1$, or

(3) $x = 1$ and $y = 0$.

This is illustrated schematically in Figure 2.1.

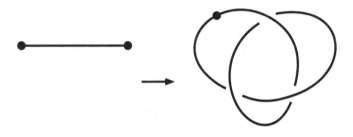

Figure 2.1

Unfortunately, with this definition the infinitely knotted loop illustrated in Figure 2.2 would be admitted into

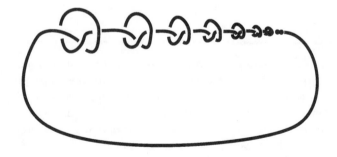

Figure 2.2

our studies. Such pathological examples are distant from the intuitive notion of a knot and the physical knotting that the theory is modelling, and so must be avoided.

Suppose for the moment that a definition similar to that indicated above were suitable. How would the idea of a deformation be captured? A natural choice would be to say that a knot J is a deformation of K if there exists a family of knots, K_t, $0 \leq t \leq 1$, with $K_0 = K$, $K_1 = J$, and with K_t close to K_s, for t close to s. Of course the idea of knots being close would have to be defined as well.

Once again, an example indicates the difficulty of using a definition based on continuity. In Figure 2.3 several steps of a deformation of a knot into an unknotted loop are illustrated. Note that at every step of the deformation the loop is a continuous simple closed curve. Somehow the definition must rule out such deformations.

One remedy is to introduce differentiability into the discussion. For instance, if the function f is required to be differentiable, with unit velocity, the possibility of a wild knot is eliminated; for the knot in Figure 2.3, the tangent is varying rapidly near the wild point where the small knots bunch up, and there is no continuous way to define a tangent direction at that wild point. Introducing differentiability into the definition of deformation is also possible, but more difficult.

An alternative solution is to use polygonal curves instead of differentiable ones. This approach avoids many technical difficulties and at the same time eliminates wild knotting, as polygonal curves are finite by nature. A theorem relating the two approaches is proved in the appendix of the text by Crowell and Fox, a good starting point for readers interested in this aspect of the theory.

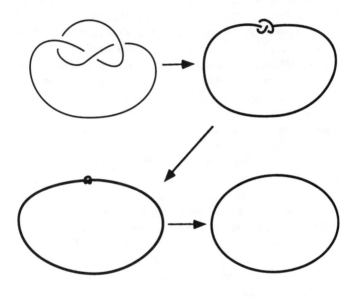

Figure 2.3

2 The Definition of a Knot

The simplest definitions in knot theory are based on polygonal curves in 3-space. Essentially a knot is defined to be a simple closed curve formed by "joining the dots."

For any two distinct points in 3-space, p and q, let $[p,q]$ denote the line segment joining them. For an ordered set of *distinct points*, $(p_1, p_2, \ldots, p_n)$, the union of the segments $[p_1, p_2]$, $[p_2, p_3], \ldots, [p_{n-1}, p_n]$, and $[p_n, p_1]$ is called a closed polygonal curve. If each segment intersects exactly two other segments, intersecting each only at an endpoint, then the curve is said to be *simple*.

☐ **DEFINITION.** *A knot is a simple closed polygonal curve in R^3.*

Figure 2.4a illustrates the simplest nontrivial knot, which is called the *trefoil*, drawn as a polygonal curve. The *unknot*, or trivial knot, is defined to be the knot determined by three noncollinear points, as illustrated in Figure 2.4b. (Note that picking a different set of three points yields a different "unknot." This ambiguity will be resolved in discussing deformations and equivalence, and in the exercises.)

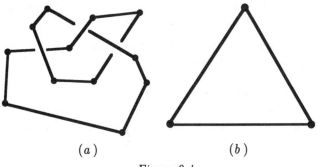

(*a*)　　　　　　　(*b*)

Figure 2.4

Knots are usually thought of, and drawn, as smooth curves and not jagged ones. An informal way of dealing with this is to view smooth knots as polygonal knots constructed from a very large number of segments. That a smooth knot can be closely approximated by a polygonal curve is intuitively clear. The formal way of dealing with this problem is to study the relationship between polygonal and differentiable knots. Knots will often be drawn smoothly in this book, but this is for aesthetic reasons, and all the figures could have been drawn polygonally instead.

There is one important observation to be made about the definition. A knot is defined to be a subset of 3-space, the union of a collection of segments. Various choices of ordered sets of points can define the same knot. For instance, cyclicly permuting the order of the points does not alter the underlying knot. Also, if three consecutive points are collinear, then eliminating the middle one does not change the underlying knot. This last observation about eliminating points along segments leads to a useful definition.

☐ **DEFINITION.** *If the ordered set* $(p_1, p_2, \ldots, p_n)$ *defines a knot, and no proper ordered subset defines the same knot, the elements of the set* $\{p_i\}$ *are called vertices of the knot.*

Finally, even if one's goal is to study only knots, links of many components will arise.

☐ **DEFINITION.** *A* link *is the finite union of disjoint knots.* (*In particular, a knot is a link with one component.*) *The* unlink *is the union of unknots all lying in a plane.*

Notice that the condition that the components of the unlink lie in a single plane is essential; examples of nontrivial links with each component unknotted have already been described. As with the definition of the unknot, ambiguities appear here; for instance, in the definition of the unlink does it matter what plane is used? Following the definition of equivalence presented in Section 3, these issues can be addressed.

EXERCISES

2.1. The ordering of the points $\{p_i\}$ used to define a knot is essential. Show that by correctly changing the ordering of the points, one might not get a knot at all. (Hint: with the vertices reordered a closed curve will still result, but is it necessarily simple?) Also, show that by changing the ordering of the points $\{p_i\}$ defining the trefoil, the resulting knot can be deformed into the unknot.

2.2. It is not clear from the definition that a knot has only one set of vertices. Prove that in fact the vertices of a knot form a well-defined set.

3 Equivalence of Knots, Deformations The next step is to give a mathematical formulation of the idea of deforming knots. This is done with the notion of equivalence, which is in turn defined via elementary deformations.

□ **DEFINITION.** *A knot J is called an elementary deformation of the knot K if one of the two knots is determined by a sequence of points $(p_1, p_2, \ldots, p_n)$ and the other is determined by the sequence $(p_0, p_1, p_2, \ldots, p_n)$, where (1) p_0 is a point which is not collinear with p_1 and p_n, and (2) the triangle spanned by (p_0, p_1, p_n) intersects the knot determined by $(p_1, p_2, \ldots, p_n)$ only in the segment $[p_1, p_n]$.*

Here a triangle is the flat surface bounded by the edges $[p_0, p_1]$, $[p_1, p_n]$, and $[p_n, p_0]$. It is defined formally as $T = \{x p_0 + y p_1 + z p_n \mid 0 \leq x, y, z, \text{ and } x + y + z = 1\}$.

The second condition in the definition assures that in performing an elementary deformation the knot does not cross itself. Figure 2.5a illustrates an elementary deformation, and 2.5b illustrates a deformation which is not permitted. As examples have indicated, such crossings can change a knot into a different type of knot. Of course, the point of the definition is to make these ideas precise.

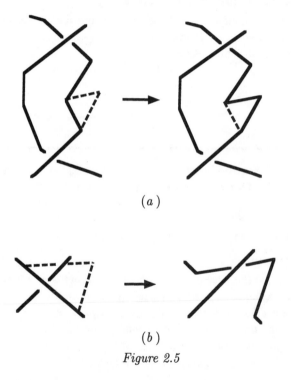

(a)

(b)

Figure 2.5

Knots K and J are called equivalent if K can be changed into J by performing a series of elementary deformations. More precisely:

☐ **DEFINITION.** *Knots K and J are called equivalent if there is a sequence of knots $K = K_0, K_1, \ldots, K_n = J$, with each K_{i+1} an elementary deformation of K_i, for i greater than 0.*

This notion of equivalence satisfies the definition of an equivalence relation; it is symmetric, transitive, and reflexive, three facts that the reader can verify.

Knot theory consists of the study of equivalence classes of knots. For instance, proving that it is impossible to deform one knot into another is the same as proving that the two knots lie in different equivalence classes. Proving that a knot is nontrivial consists of showing that it is not contained in the equivalence class of the unknot.

TERMINOLOGY
It is usual in the subject to blur the distinction between a knot and its equivalence class. For instance, rather than say that a knot is equivalent to the unknot, one just states that the knot is unknotted. Similarly, when it is said that two knots are distinct, it is meant that the knots are inequivalent. This convention seldom can cause confusion, but will be avoided in ambiguous situations.

EXERCISES
3.1. Suppose a knot lies in a plane, and bounds a convex region in that plane. (Convex means that any segment with endpoints in the region is entirely contained in the region.) Prove that the knot is equivalent to a knot with 3 vertices. That is, describe how to construct a sequence of knots, each an elementary deformation of the previous one, starting with the convex planar knot and ending with a knot having exactly 3 vertices. Hint: Apply induction on the number of vertices.

3.2. Suppose that K and J are unknots lying in the same plane. (Recall that this means that K and J are each determined by three noncollinear points.) Show that K and J are equivalent by describing a method for finding the appropriate sequence of elementary deformations.

3.3. Exercises 3.1 and 3.2 show that two convex knots in a plane determine equivalent knots. This result is true for nonconvex knots, and is called the Schonflies Theorem. Prove the Schonflies theorem for planar knots with 4 and 5 vertices.

3.4. Is every knot with exactly 4 vertices unknotted?

3.5. Let K be a knot determined by points $(p_1, p_2, \ldots, p_n)$. Show that there is a number z such that if the distance from p_1 to p_1' is less than z, then K is equivalent to the knot determined by $(p_1', p_2, \ldots, p_n)$. Similarly, show there is a z such that every vertex can be moved a distance z without changing the equivalence class of the knot. (These are both detailed arguments in epsilons and deltas.)

3.6. Prove, using 3.5, that a knot can be arbitrarily translated or rotated by a sequence of elementary deformations.

3.7. Generalize the definition of elementary deformation, and equivalence, to apply to links. (Your definition should not permit one component to pass through another.)

4 Diagrams and Projections   Although a knot is a subset of space, all our work takes place in a plane. The pictures in this book all lie on a flat piece of paper and your practice is done on a flat blackboard or piece of paper as well. How is it that a diagram on a piece of paper gives a well-defined knot? This is answered by formalizing the notion of knot diagram.

The function from 3-space to the plane which takes a triple (x,y,z) to the pair (x,y) is called the projection map. If K is a knot, the image of K under this projection is called the *projection* of K. A projection of the *figure-8* knot (knot 4_1 in the appendix) is illustrated in Figure 2.6.

It is possible that different knots can have the same projection. Once the curve is projected into the plane, it is no longer clear which portions of the knot passed over other parts. To remedy this loss of information, gaps are left in the drawings of projections to indicate which parts of the knot pass under other parts. Such

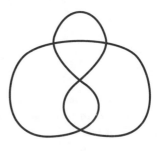

Figure 2.6

a drawing is called a knot *diagram*. In this book all the drawings of knots are really knot diagrams.

At this point the distinction between knots and equivalence classes of knots appears. Many different knots can have the same diagram, as the diagram indicates that certain portions of the knot pass over other portions, but not how high above they pass. It turns out that this does not matter! *If two knots have the same diagram they are equivalent.* To state this formally as a theorem requires a more careful study of projections.

Suppose that a knot has a projection as illustrated in Figure 2.7a. If that knot is rotated slightly in space, the resulting knot will have a projection as illustrated in

Figure 2.7b! Such knot projections have to be avoided as too much information has been lost in the projection.

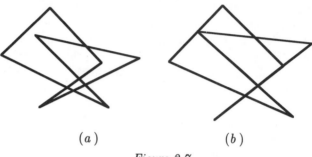

$$(a) \qquad\qquad (b)$$

Figure 2.7

☐ **DEFINITION.** *A knot projection is called a regular projection if no three points on the knot project to the same point, and no vertex projects to the same point as any other point on the knot.*

There are two theorems that make regular projections especially useful. The first states that if a knot does not have a regular projection then there is an equivalent knot nearby that does have a regular projection. The second states that if a knot does have a regular projection then all nearby knots are equivalent and also have regular projections. The notion of nearby is made precise by measuring the distance between vertices.

☐ **THEOREM 1.** *Let K be a knot determined by the ordered set of points $(p_1,\ldots,p_n)$. For every number $t > 0$ there is a knot K' determined by an ordered set $(q_1,\ldots,q_n)$ such that the distance from q_i to p_i is less than t for all i, K' is equivalent to K, and the projection of K' is regular.*

□ **THEOREM 2.** *Suppose that K is determined by the sequence* $(p_1,\ldots,p_n)$ *and has a regular projection. There is a number* $t > 0$ *such that if a knot* K' *is determined by* $(q_1,\ldots,q_n)$ *with each* q_i *within a distance of* t *of* p_i, *then* K' *is equivalent to* K *and has a regular projection.*

Knot diagrams are only defined for knots with regular projections. The theorem relating knots to diagrams is the following:

□ **THEOREM 3.** *If knots K and J have regular projections and identical diagrams, then they are equivalent.*

PROOF
One approach is the following. First arrange that K is determined by an ordered sequence $(p_1,\ldots p_n)$ and J is determined by the sequence $(q_1,\ldots,q_n)$ with the projection of p_i and q_i the same for all i. This may require introducing extra points in the defining sequences for both knots.

Next perform a sequence of elementary deformations that replace each p_i with a q_i in the defining sequence for K. These moves are first applied to all vertices which do not bound intervals whose projections contain crossing points. Finally each crossing point can be handled. □

TERMINOLOGY
A knot diagram consists of a collection of arcs in the plane. These arcs are called either *edges* or *arcs* of the diagram. The points in the diagram which correspond to double points in the projection are called *crossing points*, or just *crossings*. Above the crossing point are two segments on the knot; one is called an *overpass* or *overcrossing*, the other the *underpass* or *undercrossing*. Notice that the number of arcs is the same as the number of crossings.

With Theorem 3 it is now possible to blur the distinction between a knot and its diagram. There is usually no confusion created by not distinguishing a knot diagram from an equivalence class of a knot. To be clear, though: a knot is a subset of 3-space, knots determine equivalence classes of knots, and knots with regular projections have diagrams, which are drawings in the plane.

EXERCISES

4.1. Fill in the details of the proof of Theorem 3.

4.2. Sketch a proof of Theorem 1. (A proof can make use of Exercise 6, Section 3. A projection is regular as long as 1) no line joining two vertices is parallel to the vertical axis, 2) no vertices span a plane containing a line parallel to the vertical axis, and 3) there are no triple points in the projection. Argue that the knot can be rotated slightly to achieve conditions 1 and 2, and then deal with triple points.)

4.3. Prove Theorem 2. The previous hint should help here.

4.4. Show that the trefoil knot can be deformed so that its (nonregular) projection has exactly one multiple point.

5 Orientations   Knots can be oriented, or, informally, given a sense of direction. Recall that a knot is determined by its (ordered) set of vertices. If the ordered set of vertices is $(p_1,\ldots,p_n)$, then, as noted earlier, any cyclic permutation of the vertices gives the same knot. It is also true that reversing the order of the vertices will yield the same knot.

□ **DEFINITION.** *An oriented knot consists of a knot and an ordering of its vertices. The ordering must be chosen so that it determines the original knot. Two orderings are considered equivalent if they differ by a cyclic permutation.*

The orientation of a knot is usually represented by placing an arrow on its diagram. The connection with the definition of orientation should be clear.

The notion of equivalence is easily generalized to the oriented setting. If a knot is oriented, an elementary deformation results in a knot which is naturally oriented. Hence, an elementary deformation of an oriented knot is again an oriented knot.

□ **DEFINITION.** *Oriented knots are called oriented equivalent if there is a sequence of elementary deformations carrying one oriented knot to the other.*

One of the hardest problems that arises in knot theory is in distinguishing equivalence and oriented equivalence. The first examples of knots which are equivalent but not oriented equivalent were described by H. Trotter in 1963; for example, the (3,5,7)-pretzel knot can be oriented in two ways, and Trotter showed the resulting oriented knots are not oriented equivalent, even though they are the same when orientations are ignored.

Another related definition will be useful later.

□ **DEFINITION.** *The reverse of the oriented knot determined by the ordered set of vertices $(p_1, \ldots, p_n)$, is the oriented knot K^r with the same vertices but with their order reversed. An oriented knot K is called reversible if K and K^r are oriented equivalent. If K is not oriented, it is called reversible if for some choice of orientation it is reversible.*

EXERCISES

5.1. Formulate a definition of oriented link.

5.2. Any oriented knot, or link, determines an unoriented link. Simply ignore the orientation. Given a knot, there are at most two equivalence classes of oriented knot that determine its equivalence class, ignoring orientations. (Why?)

(a) What is the largest possible number of distinct oriented n component links which can determine the same unoriented link, up to equivalence? Try to construct an example in which this maximum is achieved. (Do not attempt to prove that the oriented links are actually inequivalent. This will have to wait until more techniques are available.)

(b) Show that any two oriented links which determine the unlink as an unoriented link are oriented equivalent.

5.3. Explain why if an unoriented knot is reversible, then for any choice of orientation it is reversible.

5.4. Show that the (p,p,q)-pretzel knot is reversible.

5.5. The knot 8_{17} is the first knot in the appendix that is not reversible, a difficult fact to prove. Find inversions for some of the knots that precede it. Several are not obvious.

5.6. Classically, what has been defined here as the reverse of a knot was called the inverse. The change in notation arose from high-dimensional considerations that will be discussed in Chapter 9. The inverse is now defined as follows. Given an oriented knot, multiplying the z-coordinates of its vertices by -1 yields a new knot, K^m, called the *mirror image*, or *obverse* of the first. The *inverse* of K is defined to be K^{mr}.

(a) How are the diagrams of a knot and its obverse and inverse related?

(b) Given a knot diagram it is possible to form a new knot diagram by reflecting the diagram through a vertical line in the plane, as illustrated in Figure 2.8. What operation on knots in 3-space does this correspond to?

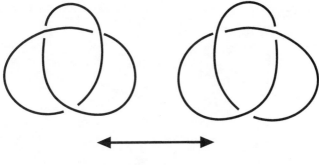

Figure 2.8

(c) Show that the operation described in part b) yields a knot equivalent to the obverse of the original knot.

CHAPTER 3:
COMBINATORIAL TECHNIQUES

The techniques of knot theory which are based on the study of knot diagrams are called combinatorial methods. These techniques are usually easy to describe and yet provide deep results. For instance, in this chapter such methods will be used to prove that nontrivial knots exist and then to demonstrate that there is in fact an infinite number of distinct knots.

Combinatorial tools often appear as unnatural or ad hoc. In many cases alternative perspectives, though more abstract, can provide insights. One of the successes of algebraic topology is to provide such perspectives, but in some cases, the efficacy of combinatorial techniques remains mysterious. Recent progress in combinatorial knot theory will be described in Chapter 10.

1 Reidemeister Moves   In what ways are diagrams of equivalent knots related? Clearly, even a single elementary deformation can have a dramatic effect on the diagram. Some of the simplest changes in a diagram that can occur when a knot is deformed are illustrated in Figure 3.1. In the figure only that portion of the diagram where a change occurs is illustrated.

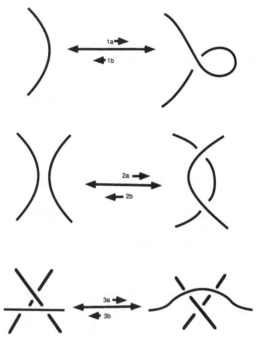

Figure 3.1

Each of the three figures represents a pair of possible changes in a diagram; each operation is paired with its inverse. These six simple operations which can be performed on a knot diagram without altering the corresponding knot are called *Reidemeister moves* . The key observation in combinatorial knot theory was made by Alexander and Briggs:

□ **THEOREM 1.** *If two knots (or links) are equivalent, their diagrams are related by a sequence of Reidemeister moves.*

PROOF

If you have already done some of the exercises showing that different diagrams can represent the same knot then this result should seem intuitively clear. In turning one diagram into the other the only changes that you ever need to make are these Reidemeister moves. The full proof is a detailed argument keeping track of a number of cases, but the main ideas are fairly simple.

Suppose that K and J represent equivalent knots, and that both have regular projections. Then K and J are related by a sequence of knots, each obtained from the next by an elementary deformation. A small rotation will assure that each knot in the sequence has a regular projection, and thus the proof is reduced to the case of knots related by a single elementary deformation.

Again after performing a slight rotation, it can be assured that the triangle along which the elementary deformation was performed projects to a triangle in the plane. That planar triangle might contain many crossings of the knot diagram. However, it can be divided up into many small triangles, each of which contains at most one crossing. This division can be used to describe the single elementary deformation in a sequence of many small elementary deformations; the effect of each on the diagram is quite simple. The proof is completed by checking that only Reidemeister moves have been applied.   □

EXERCISES

1.1. Show that the change illustrated in Figure 3.2 can be achieved by a sequence of two Reidemeister moves.

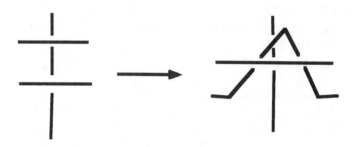

Figure 3.2

1.2. Find a sequence of Reidemeister moves that transforms the diagram of the unknot drawn in Figure 3.3. Here is a more challenging exercise: What would be the least number of Reidemeister moves needed for such a sequence? Can you prove that this is the least number that suffices?

Figure 3.3

2 Colorings   The method of distinguishing knots using the "colorability" of their diagrams was invented by Ralph Fox. The procedure is simple: A knot diagram is called *colorable* if each arc can be drawn using one of three colors, say red (R), yellow (Y), and blue (B), in such a way that 1) at least two of the colors are used, and 2) at any crossing at which two colors appear, all three appear. Figure 3.4 illustrates a coloring of a knot diagram. Exercise 2.1 is a quick problem, asking you to check which of the diagrams for knots with 7 or fewer crossings, as illustrated in Appendix 1, are colorable.

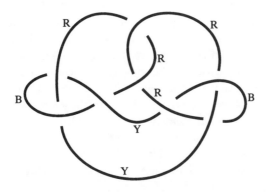

Figure 3.4

Is it possible that some diagrams for a knot are colorable while others are not? Our first result in combinatorial knot theory is that the answer is no.

☐ **THEOREM 2.** *If a diagram of a knot, K, is colorable, then every diagram of K is colorable.*

Hence the following definition makes sense:

☐ **DEFINITION.** *A knot is called colorable if its diagrams are colorable.*

The proof of Theorem 2 is the model for most of the proofs of later combinatorial results. But before giving it, one immediate consequence should be noted; nontrivial knots exist! Clearly the unknot is not colorable because its standard projection cannot be colored. It follows that any colorable knot is nontrivial. Further consequences appear in the exercises.

PROOF

(Theorem 2) It is sufficient to show that if a Reidemeister move is performed on the colorable diagram of a knot, then the resulting diagram is again colorable. Hence, the proof breaks into six steps, one for each Reidemeister move. Each step consists of checking various cases and none is difficult, although some are a bit tedious. One step is presented here; the others are left to the exercises.

Suppose that Reidemeister move 2b is performed on a colored knot diagram. It must be shown that the new diagram is again colorable. There are two cases. In the first, the arcs are colored with two (and hence three) colors, as illustrated in Figure 3.5a. (Only the affected portions of the knots are included in these illustrations.) The new diagram can be colored as before, with the altered section colored as in Figure 3.5b. As two colors still appear, the resulting diagram is still colorable.

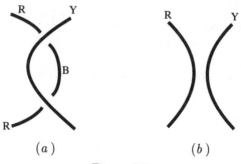

Figure 3.5

The second possibility is that both of the affected arcs start out colored with the same color. In this case, after performing the Reidemeister move the arcs can still be colored with that same color and the rest of the diagram can be colored as it was originally. All the requirements of colorability are still satisfied.

Checking Reidemeister moves 1a, 1b, and 2a, are all as simple as this. Moves 3a and 3b present a few more cases to check.   □

EXERCISES

2.1. Which of the knot diagrams with seven or fewer crossings, as illustrated in Appendix 1, are colorable?

2.2. For which integers n is the $(2,n)$-torus knot in Figure 3.6a colorable? The knot illustrated in Figure 3.6b is called the *n-twisted double of the unknot*, where n denotes the number of twists in the vertical band. The trefoil results when $n = 1$. What if $n = -1$? For which values of n is the n-twisted double of the unknot colorable?

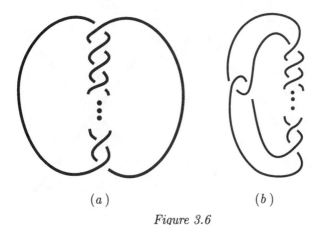

(a) (b)

Figure 3.6

2.3. Discuss the colorability of the (p,q,r)-pretzel knots.

2.4. (a) Prove the coloring theorem for Reidemeister move 1a.

(b) How many cases need to be considered in proving Theorem 1 for Reidemeister move 3a?

(c) Check each of these cases.

(d) Complete the proof of Theorem 1.

2.5. Given an oriented link of two components, J and K, it is possible to define the *linking number* of the components as follows. Each crossing point in the diagram is assigned a sign, $+1$ if the crossing is right-handed and -1 if it is left-handed. (A *right-handed* crossing is a crossing at which an observer on the overcrossing, facing in the direction of the overcrossing, would view the undercrossing as passing from right to left. Right and left crossings are illustrated in Figure 3.7.) The linking number of K and J, $\ell k(K,J)$, is defined to be the sum of the signs of the crossing points where J and K meet, divided by 2.

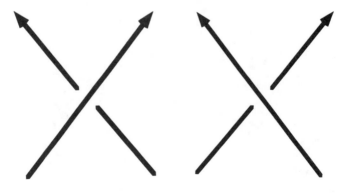

Figure 3.7

(a) Use the Reidemeister moves to prove that the linking number depends only on the oriented link, and not on the diagram used to compute it.

(b) Figure 3.8 illustrates an oriented *Whitehead link*. Check that it has linking number 0.

(c) Construct examples of links with different linking numbers.

2.6. This exercise demonstrates that the linking number is always an integer. First note that the sum used to compute linking numbers can be split into the sum of the signs of the crossings where K passes over J, and the sum of the crossings where J passes over K.

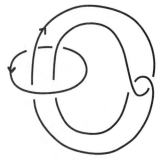

Figure 3.8

(a) Use Reidemeister moves to prove that each sum is unchanged by a deformation.

(b) Show that the difference of the two sums is unchanged if a crossing is changed in the diagram.

(c) Show that if the crossings are changed so that K always passes over J, the difference of the sums is 0. (This link can be deformed so that K and J have disjoint projections.)

(d) Argue that the linking number is always an integer, given by either of the two sums. (This is the usual definition of linking number. The definition in Exercise 2.5 makes it clear that $\ell k(K,J) = \ell k(J,K)$.)

2.7. The definition of colorability is often stated slightly differently. The requirement that at least two colors are used is replaced with the condition that all three colors appear.

(a) Show that the unlink of two components has a diagram which is colorable using all three colors and another diagram which is colorable with exactly two colors.

(b) Why is it true that for a knot, once two colors appear all three must be used, whereas the same statement fails for links?

(c) Explain why the proof of Theorem 2 applies to links as well as to knots.

2.8. Prove that the Whitehead link illustrated in Figure 3.8 is nontrivial, by arguing that it is not colorable.

2.9. In this exercise you will prove the existence of an infinite number of distinct knots by counting the number of colorings a knot has.

If a knot is colorable there are many different ways to color it. For instance, arcs that were colored red can be changed to yellow, yellow arcs changed to blue, and blue arcs to red. The requirements of the definition of colorability will still hold. There are six permutations of the set of three colors, so any coloring yields a total of six

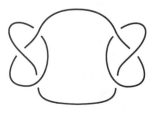

Figure 3.9

colorings. For some knots there are more possibilities.

(a) Show that the standard diagram for the trefoil knot has exactly six colorings.

(b) How many colorings does the *square knot* shown in Figure 3.9 have?

(c) The number of colorings of a knot projection depends only on the knot; that is, all diagrams of a knot will have the same number of colorings. Outline a proof of this.

(d) Use the connected sum of n trefoils, illustrated in Figure 3.10, to show that there are an infinite number of distinct knots.

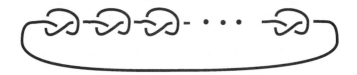

Figure 3.10

3 A Generalization of Colorability, mod p Labelings   How can colorability be generalized? Is it possible to use more than three colors to describe new methods of distinguishing knots? There are actually several ways to generalize colorability, the first of which is presented in this section.

In describing the method of colorings in the previous section, instead of labeling the arcs of the knot diagram with colors, three integers, 0,1, and 2, could have been used. The condition on colorings at crossings translates into the simple statement that if the overcrossing is labeled x and the two other arcs y and z, then the difference $2x - y - z$ is divisible by 3, or, more succinctly, $2x - y - z = 0 \pmod 3$. (Check that this condition is equivalent to the coloring condition.) A possible generalization immediately appears:

☐ **DEFINITION.** *A knot diagram can be labeled mod p if each edge can be labeled with an integer from 0 to $p-1$ such that 1) at each crossing the relation $2x - y - z = 0 \pmod p$ holds, where x is the label on the overcrossing and y and z the other two labels, and 2) at least two labels are distinct.*

Figure 3.11 illustrates a mod 7 labeling of a knot.

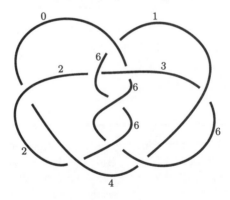

Figure 3.11

For reasons that will be made clear in the exercises, p will be restricted to the odd primes. In Exercise 3.3 the reader is invited to check that whether or not a knot diagram can be labeled mod p depends only on the equivalence class of the knot, so that Theorem 2 generalizes to this new situation. Figure 3.12 illustrates one step; if Reidemeister move 2b is performed on a labeled diagram, the resulting diagram can again be labeled.

Figure 3.12

□ **THEOREM 3.**   (*Labeling theorem*) *If some diagram for a knot can be labeled* mod p *then every diagram for that knot can be labeled* mod p.

EXERCISES

3.1. Determine which knots with 6 or fewer crossings can be labeled mod 5.

3.2. For what primes p can the trefoil knot diagram be labeled mod p?

3.3. Prove Theorem 3 by showing that if any Reidemeister move is performed on a labeled diagram, the resulting diagram can again be labeled.

3.4. Show that if all the labels of a knot that is labeled mod 3 are multiplied by 5, the resulting labeling is a labeling mod 15. This gives some indication as to why p is restricted to the primes.

3.5. If p is 2, other difficulties come up. Explain why no knot can be labeled mod 2. (Modulo 2, what does the crossing relationship say?)

3.6. Check that the theory of labelings applies to links of many components.

3.7. Show that the knots 4_1, 7_1, and 8_{16} are distinct by using mod 5 and mod 7 labelings. (Find mod 5 and mod 7 labelings of 8_{16}.)

---

4 Matrices,
Labelings, and
Determinants

Linear algebra simplifies the problem of labeling knot diagrams; just as important is the fact that, with the introduction of matrices, many new knot invariants appear. Some of these invariants are introduced here. These invariants are studied in greater depth in Chapter 7.

Here is an algebraic reduction of the problem. Given a knot diagram, label each arc of the diagram with a variable, say x_i. At each crossing a relation between the variables is defined: if arc x_i crosses over arcs x_j and x_k, then $2x_i - x_j - x_k = 0 \pmod{p}$. A knot can be labeled mod p if there is a mod p solution to this system of equations with not all x_i equal.

Whether or not a knot is colorable, or can be labeled mod p, has now been reduced to a problem of linear algebra, that of studying the solutions to a system of linear equations. As usual in linear algebra, the use of matrices will simplify the problem.

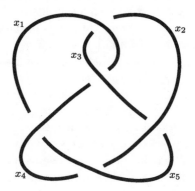

Figure 3.13

For example, the knot in Figure 3.13 is drawn with its arcs labeled and its crossings numbered. The corresponding system of equations that needs to be solved is given by the matrix below. The rows correspond to the equations determined by each crossing, the columns to the variables taken in order.

$$\begin{pmatrix} 2 & -1 & -1 & 0 & 0 \\ -1 & 0 & 2 & -1 & 0 \\ -1 & 0 & 0 & 2 & -1 \\ 0 & -1 & 0 & -1 & 2 \\ 0 & 2 & -1 & 0 & -1 \end{pmatrix}$$

Standard techniques of linear algebra apply to solving systems of equations mod p as well as for finding real or rational solutions. (Formally, for p prime the integers mod p form a field.) Unfortunately, the added condition in the present problem, that the solutions have at least two of the x_i distinct, introduces a few subtleties that need to be addressed before general results can be presented.

Two preliminary observations are needed. First note that setting each $x_i = 1$ yields a solution to the system of equations. Second, observe that any two solutions can be added together to yield another solution.

These remarks imply that if there is a solution with not all entries equal, there is such a solution with $x_n = 0$. (x_n could be replaced with any other x_i here.) Conversely, a *nontrivial* solution with $x_n = 0$ results in a labeling of the knot. Hence, a solution with not all x_i equal corresponds to a nontrivial solution to the system of equations determined by the original matrix with its last column deleted.

It is easier to work with problems related to square matrices, and fortunately the given problem can be reduced to this setting. This is done by showing that any one of the equations is a consequence of the others. In terms of the matrix, multiplying certain of the rows by -1 results in a matrix with its rows adding to 0.

The correct choice of -1's is not obvious; here is the algorithm: Orient the knot. At each crossing in the diagram put a dot to the right of the overcrossing, just before the crossing point. Now, count how many arcs of the diagram must be crossed by a path from the dot to a point in the plane far from the diagram. If an odd num-

Figure 3.14

ber of arcs are crossed, then multiply the corresponding row of the matrix by -1. It is fairly simple to show that the sum of the rows is now trivial. In Figure 3.14 the crossings that correspond to rows that are multiplied by -1 are marked.

The following result summarizes the discussion above.

□ **THEOREM 4.** *There is an $n \times n$ matrix corresponding to a knot diagram with n arcs. Deleting any one column and any one row yields a new matrix. The knot can be labeled* mod p *if and only if the corresponding set of equations has a nontrivial* mod p *solution.*

Of course whether or not the system of equations has a nontrivial solution depends on the determinant of the matrix. A solution exists if the determinant is 0, or, working mod p, if the determinant is divisible by p. Furthermore, the number of solutions is determined by the mod p nullity of the matrix.

(The nullity of a matrix is the dimension of the kernel of the matrix, thought of as a linear transformation. More algorithmically, any square matrix with entries in a field, (mod p entries in the present case), can be diagonalized by performing row and column operations; that is, by adding multiples of rows or columns to other rows or columns respectively. The number of 0's on the diagonal (or entries divisible by p if working mod p) is the *nullity*. With more care, a square integer matrix can be diagonalized, using only integer row and column operations. Performing this integer diagonalization performs mod p diagonalizations for all p simultaneously. The exercises illustrate these procedures.)

□ **DEFINITION.** *The determinant of a knot is the absolute value of the determinant of the associated $(n-1) \times (n-1)$ matrix constructed above.*

□ **DEFINITION.**   *The* mod *p rank of a knot is the* mod *p nullity of the associated* $(n-1) \times (n-1)$ *matrix constructed above.*

Of course, for these two definitions to give well-defined invariants, it must be proved that none of the choices involved, of either the knot diagram or the ordering of the labels on the arcs and crossings, affects the determinant or mod p rank of the associated matrix.

□ **THEOREM 5.**   *The determinant of a knot and its* mod *p rank are independent of the choice of diagram and labeling.*

PROOF

There are two parts to the proof. The first is purely linear algebra, observing facts about the determinant and nullity of matrices. The second calculates the effect of the choice of labelings and the Reidemeister moves on the associated matrix.

As far as the linear algebra goes, a needed result states that if, for a square matrix, the sum of the rows and the sum of the columns is 0, then if a row and column are removed, the nullity (and the absolute value of the determinant) of the resulting matrix does not depend on which row and column were removed. A simpler result states that if the matrix is changed by adding a new row and column, each containing all 0's except for a single 1 on the diagonal, then the nullity and determinant are unaffected.

The rest of the argument checks the effect of the Reidemeister moves on the associated matrix. For example, Reidemeister move 2a introduces two new rows and two new columns. Two of the new columns result from splitting one of the old arcs into two, and hence the sum of

those two columns has entries determined by the one old
column. A few row and column operations show that the
new matrix can be changed into the old, with two new
rows and columns added, each of which has a single ± 1 in
it. The full argument for this and the other Reidemeister
moves, is left to the reader.   □

TORSION INVARIANTS

The determinant and ranks are captured by stronger invariants. It is relatively easy to diagonalize a matrix when
working mod p; any nonzero entry can be used to clear
out a row and column. Diagonalizing over the integers is
harder, though possible, as is proved in most modern algebra texts in the classification of abelian groups. The proof
uses the Euclidean algorithm. The typical result states
that a square integer matrix can be diagonalized so that
each entry on the diagonal divides the next entry. If the
matrix associated to a knot is diagonalized in this way, the
resulting diagonal entries are called the *torsion invariants*
of the knot. Their product is the determinant of the knot,
and the number of entries which are divisible by p is the
mod p rank of the knot.

The proof that these are well-defined knot invariants
will not be given. The best approach relies on the theory
of abelian groups. The matrix associated to a knot can be
viewed as a presentation matrix for an abelian group. The
various alterations in the matrix do not affect the group so
determined, and the torsion invariants are just the torsion
invariants of this group.

EXERCISES

4.1. For each knot with 6 or fewer crossings find the associated matrix, and its determinant. In each case, for what
p is there a mod p labeling?

4.2. The knots 8_{18} and 9_{24} both have determinant 45. Check that one has a mod 3 rank of 1, while the other has a mod 3 rank of 2. The knots 8_8 and 9_{49} both have determinant 25. Compute their mod 5 ranks.

4.3. Prove the linear algebra results stated in the proof of Theorem 5.

4.4. Because the unknot has particularly simple diagrams, the arguments given above really need to be modified slightly. The two diagrams for the unknot that cause difficulties are the diagram with no crossings, and the diagram with exactly one crossing. What goes wrong in these cases? Why don't these problems occur in other situations? How would you correct for these minor problems? (Define the determinant and nullity of a 0×0 matrix to be 1.)

4.5. Prove that the determinant of a knot is always odd. (See Exercise 5 of the previous section, relating to mod 2 labelings. Also, this result does not apply for links of more than one component.)

4.6. Show that if a knot has mod p rank n, then the number of mod p labelings is $p(p^n - 1)$.

5 The Alexander Polynomial

In the previous section it was seen that the simple notion of colorability leads to a study of determinants of matrices. The following description of the Alexander polynomial greatly extends the use of matrices and determinants. In this case, rather than work with entries that are integers the entries of the matrix are polynomials.

Alexander's original description was based on labeling the regions in the plane bounded by the arcs of the diagram, and Reidemeister was the first to give a presentation focusing on the arcs. Since then, many alternative definitions have been found. Chapter 10 provides a modern viewpoint, one that is quite simple, and that provides access to many new invariants.

To compute the Alexander polynomial of a knot, $A_K(t)$, first pick an oriented diagram for K. Number the arcs of the diagram, and separately number the crossings. Next, define an $n \times n$ matrix, where n is the number of crossings (and arcs) in the diagram, according to the following procedure:

If the crossing numbered ℓ is right-handed with arc i passing over arcs j and k, as illustrated in Figure 3.15a, enter a $1-t$ in column i of row ℓ, enter a -1 in column j of that row, and enter a t in column k of the row. If the crossing is left-handed, as illustrated in Figure 3.15b, enter a $1-t$ in column i of row ℓ, enter a t in column j and enter -1 in column k of row ℓ. All of the remaining entries of row ℓ are 0. (An exceptional case occurs if any of i, j, or k are equal. In this exceptional case, the sum of the entries described above is put in the appropriate column. For instance, if $j = k$ for some left-handed crossing, enter $-1+t$ in column j. What if $j = k$ at a right-handed crossing?)

□ **DEFINITION.** *The $(n-1) \times (n-1)$ matrix obtained by removing the last row and column from the $n \times n$ matrix just described is called an Alexander matrix of K. The determinant of the Alexander matrix is called the Alexander polynomial of K. (The determinant of a 0×0 matrix is defined to be 1.)*

Unfortunately, this polynomial depends on the choice of the original diagram as well as on the other choices involved in its description. That dependence is captured by the following theorem.

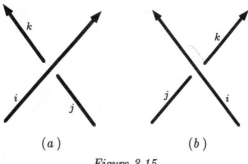

Figure 3.15

☐ **THEOREM 6.** *If the Alexander polynomial for a knot is computed using two different sets of choices for diagrams and labelings, the two polynomials will differ by a multiple of $\pm t^k$, for some integer k.*

For example, applying this procedure to the trefoil yields the polynomial $t^2 - t + 1$. Another set of choices might give $-t^4 + t^3 - t^2$. See below.

SKETCH OF PROOF
The argument is more detailed than, but quite similar to, the proof of Theorem 5. With some care, the reader should be able to check the effect of performing Reidemeister moves on the Alexander matrix. The complete proof includes one new difficult step; analyzing the effect of a change of orientation. It will be shown that the Alexander polynomial of the reverse of a knot K is obtained from

the Alexander polynomial of K by substituting t^{-1} for t and multiplying by an appropriate power of t, and perhaps multiplying by -1. (See Exercise 5.7.) Hence, the independence of the Alexander polynomial on orientation follows from its symmetry; replacing t with t^{-1} returns the same polynomial multiplied by some power of t. This symmetry property will be discussed in Chapter 6. (Alexander was unable to find a proof; a complete argument was first given by Seifert.)

EXAMPLES

The trefoil knot provides the simplest example of a knot with nontrivial Alexander polynomial. Figure 3.16 indicates a labeling of the arcs and crossings. The associated matrix is:

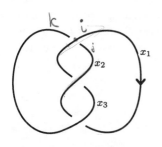

Figure 3.16

$$\begin{pmatrix} 1-t & -1 & t \\ t & 1-t & -1 \\ -1 & t & 1-t \end{pmatrix}$$

Deleting the bottom row and the last column gives a 2×2 Alexander matrix with determinant $t^2 - t + 1$.

Consider a harder example, the $(2,n)$-torus knot, shown in Figure 3.17. If the diagram is labeled as was done for the trefoil, the Alexander polynomial is given as the determinant of the $(n-1) \times (n-1)$ matrix

$$\begin{pmatrix} 1-t & -1 & 0 & 0 & \cdots & 0 \\ t & 1-t & -1 & 0 & \cdots & 0 \\ 0 & t & 1-t & -1 & \cdots & 0 \\ & & & \vdots & 1-t & -1 \\ 0 & \cdots & \cdots & 0 & t & 1-t \end{pmatrix}$$

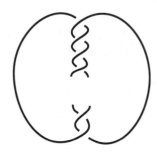

Figure 3.17

Clearly, to compute the exact determinant here would take a fairly detailed inductive argument. (The result turns out to be $(t^n + 1)/(t + 1)$.) Without actually computing the determinant it is easily proved that for different positive n the Alexander polynomials are distinct. Note first that the coefficient of the lowest degree term, the constant term, is the determinant of the matrix obtained by setting $t = 0$. The result is 1. The highest degree term is found by taking the determinant of the matrix containing only the t terms of the matrix above; that is remove all the $\pm 1's$. The resulting determinant is t^{n-1}.

Hence the Alexander polynomial of the $(2,n)$-torus knot is of degree exactly $n - 1$. In particular, these knots form an infinite family of distinct knots, all of which are distinguished by the Alexander polynomial.

EXERCISES

5.1. Compute the Alexander polynomial for several knots in the appendix.

5.2. Relate the value of the Alexander polynomial of a knot evaluated at -1 to the determinant of the knot, defined in the previous section.

5.3. Check that Reidemeister move 1a does not change the Alexander polynomial.

5.4. It is possible to construct knots with the same polynomial, but which can be distinguished by their mod p ranks for some p. Compute the polynomials of 8_{18} and 9_{24}

to check that they are identical. In Exercise 4.3 of this chapter these knots were distinguished using the mod 3 ranks.

5.5. Show that the knot in Figure 3.18 has Alexander polynomial 1. (This is one of only two knots with 11 or fewer crossings that has trivial polynomial, other than the unknot.) Use Exercise 5.2 to argue that the knot cannot be distinguished from the unknot using labelings. Stronger algebraic techniques (Chapter 5) or combi-

Figure 3.18

natorial tools (Chapter 10) can be used to prove it is nontrivial.

5.6. Prove that a knot and its mirror image, as illustrated in Figure 3.19, have the same polynomial. (Hint: Label the mirror image in the obvious way, but reverse its orientation.)

Figure 3.19

5.7. Show that the Alexander polynomial of K with its orientation reversed is obtained from the polynomial of K by substituting t^{-1} for t, and multiplying by the appropriate power of t, and perhaps changing sign.

CHAPTER 4:
GEOMETRIC TECHNIQUES

Consider the surface drawn in Figure 4.1. It is built from a disk by attaching two twisted bands. Note that the boundary, or edge, of the surface is a knotted curve. In fact, the boundary is a trefoil knot.

Figure 4.1

By studying the surface it is possible to learn more about the trefoil knot. In general, the term *geometric techniques* refers to the methods of knot theory that are based on working with surfaces. The use of these methods is motivated by a theorem stating that for every knot there is some surface having that knot as its boundary. An important application, on which this chapter ends, is the prime decomposition theorem for knots.

The first section of this chapter presents the basic definition of surface. The discussion corresponds closely to that of Chapter 2 where knot is defined. Naturally the definition is more technical. For a knot the interest is entirely in its placement in space; a surface has additional structure which is independent of its placement. For instance, the surface in Figure 4.1 is clearly different from a disk. The concept of internal, or *intrinsic*, properties of surfaces is made precise with the notion of homeomorphism, that is also described in Section 1.

Section 2 presents the fundamental theorems concerning surfaces. These results completely classify surfaces in terms of intrinsic properties. Once this internal structure of surfaces is understood the focus can shift to the placement of surfaces in space and to the knotted boundaries of surfaces. Section 3 begins the application of surface theory to knot theory; it is proved that every knot is the boundary of some surface. Sections 4 and 5 address the prime decompostion theorem, with Section 4 devoted to building the tools of the proof and Section 5 outlining the details of the argument.

1 Surfaces and Homeomorphisms

As with knots, it is possible to define a surface using the notion of differentiability. Again, a simpler working definition can be given using polyhedra.

Any 3 noncollinear points in 3-space, p_1, p_2, and p_3, form the vertices of a unique triangle. That triangle is defined to be the set of points

$$\{xp_1 + yp_2 + zp_3 \mid x + y + z = 1,\ x,\ y,\ z \geq 0\},$$

where each p_i is thought of as a vector in R^3. The union of a finite collection of triangles is called a *polyhedral surface* if: (1) each pair of triangles is either disjoint or their intersection is a common edge or vertex, (2) at most two triangles share a common edge, and (3) the union of the edges that are contained in exactly one triangle is a disjoint collection of simple polygonal curves, called the *boundary* of the surface. This third condition rules out such possibilities as a surface being the union of exactly two triangles meeting at a vertex. (In this case the union of the edges contained in exactly one triangle would be all six edges; these form two unknots meeting in the common vertex—they are not disjoint.) Figure 4.2 illustrates a simple polyhedral surface, a planar square with a square hole in its center. It is illustrated as the union of a collection of triangles.

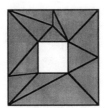

Figure 4.2

Surfaces will be drawn smoothly. Any smooth surface can be closely approximated by a polyhedral surface, but as the number of triangles required can be extremely large, it is easier to leave that *triangulation* out of the illustration. The details of the relationship between smooth and polyhedral surfaces is part of the foundational material of geometric topology.

ORIENTATION

The intuitive approach to orientability states that a surface is orientable if it is two-sided. The Möbius band is the standard example of a nonorientable surface. In calculus, a surface is called orientable if there is a nowhere vanishing vector field normal to the surface. For polyhedral surfaces there is a simple definition which corresponds to both the intuitive idea and the formal definition given in calculus.

☐ **DEFINITION.** *A polyhedral surface is orientable if it is possible to orient the boundary of each of its constituent triangles in such a way that when two triangles meet along an edge, the two induced orientations of that edge run in opposite directions.*

A surface can be *triangulated*, that is, described as the union of triangles, in many different ways, and the definition of orientability appears to depend on the choice of triangulation. However, whether or not a surface can be oriented is actually independent of the choice of triangulation.

HOMEOMORPHISM

A notion of deformation of polyhedral surfaces can be given in much the same way as was done for knots. An important observation is that, although one surface might not be deformable into a second surface, the two might be intrinsically the same; that is, they are indistinguishable without reference to how they sit in space. For example, the number of boundary components of a surface is intrinsic; an inhabitant of the surface could determine this number.

However, whether or not the boundary is knotted can only be seen from a three-dimensional perspective.

This idea of intrinsic equivalence is formally defined as *homeomorphism.* Surfaces F and G in 3-space are called homeomorphic if there is a continuous function with domain F and range G which is both one-to-one, and onto. For polyhedral surfaces there is an alternative definition. Note that there are many ways that a triangle can be subdivided into smaller triangles; a few such subdivisions are illustrated in Figure 4.3. Triangulations of surfaces can similarly be subdivided so as to yield finer triangulations.

Figure 4.3

☐ **DEFINITION.** *Polyhedral surfaces are called homeomorphic if, after some subdivision of the triangulations of each, there is a bijection between their vertices such that when three vertices in one surface bound a triangle the corresponding three vertices in the second surface also bound a triangle.*

Determining whether or not two surfaces are homeomorphic can be difficult. It might first come as a sur-

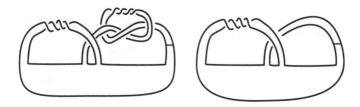

Figure 4.4

prise that the surfaces illustrated in Figure 4.4 are homeomorphic. (In the illustrations surfaces will usually not be
shaded any more.)

A homeomorphism from one to the other is given by
the map that cuts the first along the dotted line, unknots
and untwists the band, and then reattaches it. The map
is easily seen to be one-to-one and onto. Continuity follows from the fact that points that are close together on
the original band are mapped to close points on the image
band. Notice that this homeomorphism does not preserve
the knot type of the boundary! In a case such as this it
would be extremely complicated to write the map down ex-
plicitly in terms of coordinates. Triangulating the surfaces
and finding the bijection would be completely unmanage-
able. In the next section tools are developed that greatly
simplify the use of surfaces.

EXERCISES
1.1. Show that the boundary of the surface illustrated in
Figure 4.1 is the trefoil knot.
1.2. The surface in 4.1 is homeomorphic to the same surface with the bands untwisted. Why? By comparing their

boundaries, show that the surface with its bands twisted cannot be deformed into the one with untwisted bands.

1.3. Given a knot diagram, it is possible to construct a surface by "checkerboarding" the plane. Figure 4.5 shows this for two diagrams of the trefoil. Each surface was constructed by darkening in alternate regions of the plane determined by the knot projection. The first surface in 4.5 is nonorientable. (If you start on the top of the surface and travel around it once, you have gone through three twists, and hence finish on the other side.) The other surface is orientable. Redraw it using two colors to distinguish the two sides. Which of the diagrams for knots of 7 or fewer crossings in the Appendix result in orientable surfaces when checkerboarded?

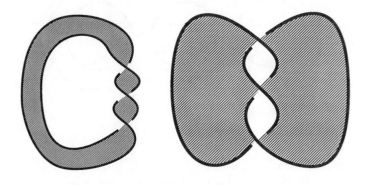

Figure 4.5

2 The Classification of Surfaces

Several connected orientable surfaces without boundary are illustrated in Figure 4.6. Associated to these surfaces is an integer called the genus of the surface, which roughly counts the number of holes. It turns out that for *any* oriented surface there is an associated number called the genus.

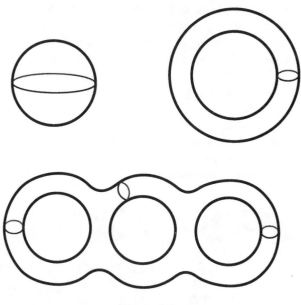

Figure 4.6

A theorem, called *the classification of surfaces*, implies that connected oriented surfaces *without boundary* are homeomorphic if and only if the they have the same genus. (Recall once again that homeomorphic surfaces need not be deformable into each other in 3-space.) A more general classification of surfaces applies to surfaces with boundary.

EULER CHARACTERISTIC AND GENUS
The Euler characteristic is an easily defined invariant of a polyhedral surface. Its definition is stated in terms of

a specific triangulation, and a basic result, usually proved
using algebraic topology, says that its value is independent
of choice of triangulations. Consequently, the Euler characteristics of homeomorphic surfaces are equal. The Euler
characteristic and genus are difficult to compute from the
definitions alone. The following results greatly simplify
their calculation.

☐ **DEFINITION.** *If a polyhedral surface S is triangulated
with F triangles, and there are a total of E edges and V
vertices in the triangulation, then the Euler characteristic
is given by* $\chi(S) = F - E + V$.

For example, in the octahedron illustrated below,
there are 8 faces, 12 edges, and 6 vertices. Therefore its
Euler characteristic is $8 - 12 + 6 = 2$.

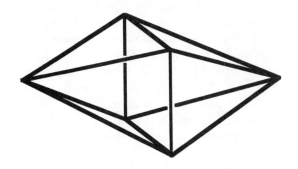

Figure 4.7

The genus of a surface is defined in terms of its Euler
characteristic. Initially, the definition appears to introduce
unnecessary algebra, but many simplifications will derive
from it.

□ **DEFINITION.** *The genus of a connected orientable surface S is given by*

$$g(S) = \frac{2 - \chi(S) - B}{2},$$

where B is the number of boundary components of the surface.

□ **THEOREM 1.** *If two surfaces intersect in a collection of arcs contained in their boundary, the Euler characteristic of the union is the sum of their individual Euler characteristics minus the number of arcs of intersection.*

PROOF

The basic idea of the proof is simple. Suppose that each arc of intersection is a single edge of a triangle on each surface. Then the triangulations of the surfaces piece together to give a triangulation of the union. The count that is used to compute the Euler characteristic of each surface separately gets a contribution of 1 from each edge of intersection (-1 for the edge, and $+2$ for its endpoints.) Hence for the sum of the two Euler characteristics there is a contribution of $+2$ from each edge of intersection. However, in the union there is a contribution of only $+1$ from each edge. The result follows.

If each arc is not a single edge of a triangle, it can be arranged to be the union of edges, after subdividing. Again it turns out that the contribution of each arc toward the total Euler characteristic is $+1$, and the rest of the argument is the same.   □

EXAMPLE

Many of the surfaces that arise are formed as disks with twisted bands added. (See Figures 4.1 and 4.4.) As the Euler characteristic of a disk is 1 (compute it for a single triangle), and a band is just an elongated disk, the Euler characteristic of a single disk with bands added is

$$(1 + \#(\text{bands})) - 2(\#(\text{bands})) = 1 - \#\text{bands}.$$

(Each band contributes two arcs of intersection.) If the surface is formed by adding bands to a collection of disjoint disks, the resulting surface has Euler characteristic $(\#\text{disks}) - (\#\text{bands})$.

☐ **COROLLARY 2.** *If two connected orientable surfaces intersect in a single arc contained in each of their boundaries, the genus of the union of the two surfaces is the sum of the genus of each.*

PROOF

Express the Euler characteristic in terms of the genus and apply Theorem 1. Note that one boundary component is lost in forming the union. Exercise 4.3 asks for the details. ☐

Theorem 3 follows from a calculation similar to that of Theorem 1:

☐ **THEOREM 3.** *If a connected orientable surface is formed by attaching bands to a collection of disks, then the genus of the resulting surface is given by*

$$(2 - \#\text{disks} + \#\text{bands} - \#\text{boundary components})/2.$$

One more result of this sort will be needed later on.

□ **THEOREM 4.** *If two surfaces intersect in a collection of circles contained in the boundary of each, the Euler characteristic of their union is the sum of their Euler characteristics.*

PROOF

The argument is similar to that of Theorem 1. In computing the Euler characteristic of a surface, each boundary component contains an equal number of edges and vertices of the triangulation. Hence, it contributes 0 to the total Euler characteristic. The same is true for the union.   □

CLASSIFICATION THEOREMS

In knot theory the main interest in surfaces concerns those with boundary. Hence, the statements of the classification theorems are restricted to this setting. The first part of the classification gives a family of standard models for surfaces. The second gives the homeomorphism classification of these models.

□ **THEOREM 5.** (*Classification I*) *Every connected surface with boundary is homeomorphic to a surface constructed by attaching bands to a disk.*

SKETCH OF PROOF

The proof of this theorem is technical, and the details appear in the references. Here is the overall idea. Fix a triangulation of the surface. A small neighborhood of each vertex forms a disk. Thin neighborhoods of the edges form bands joining the disks together. Hence, a neighborhood of the edges is homeomorphic to a union of disks with bands added. Two steps remain. The more difficult one shows that adding the faces has the same effect as not attaching certain of the bands. The other one shows that the number of disks can be reduced to one, and is detailed in the exercises.   □

□ **THEOREM 6.** (*Classification II*) *Two disks with bands attached are homeomorphic if and only if the following three conditions are met:*

(1) *they have the same number of bands,*
(2) *they have the same number of boundary components,*
(3) *both are orientable or both are nonorientable.*

EXAMPLE
The surface in Figure 4.8a consists of two disks joined together by three twisted bands. The boundary is the $(5, -3, 7)$-pretzel knot. If that surface is deformed by pushing in a narrow strip through the center band, the resulting surface can be further deformed to appear as in Figure 4.8b.

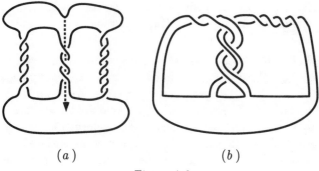

(a)   (b)

Figure 4.8

EXERCISES

2.1. Use Theorem 3 to compute the genus of the surface illustrated in Figure 4.9 below.

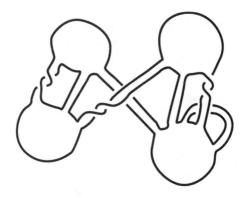

Figure 4.9

2.2. Provide the details of the proof of Theorem 3.

2.3. Prove Corollary 2.

2.4. Use Theorem 5 to prove that the only genus 0 surface with a single boundary component is the disk.

2.5. Generalize the construction illustrated in Figure 4.8 to arbitrary pretzel knots. For what values of p, q, and r, is the surface orientable?

2.6. By the classification of surfaces, the punctured torus in Figure 4.10a can be deformed into a disk with bands attached. Find a deformation into the disk with two bands illustrated in Figure 4.10b. (The punctured torus has a subsurface, which is outlined. Your deformation should consist of two steps. First, deform the entire surface onto the subsurface; then, deform the subsurface to appear as the disk with bands added.)

2.7. If a surface consists of two disks with a single band joining them, it is homeomorphic to a single disk with no bands attached. Based on such an observation argue that any connected surface which is built by adding bands to a collection of disks can in fact be built starting with only one disk. (This observation is of practical importance: The surfaces that knots bound will initially be constructed from several disks. Calculations of knot invariants coming from surfaces are much easier if the surface is described using only one disk.)

2.8. Prove that the genus of a surface is nonnegative by using induction on the number of bands.

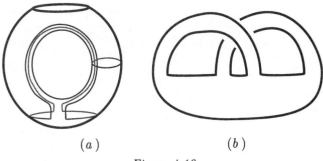

(a) (b)

Figure 4.10

2.9. Prove that the genus of an orientable surface is an integer. (Apply induction on the number of bands, and check the effect of adding an (oriented) band on the number of boundary components.

2.10. Prove that every connected orientable surface is homeomorphic to a surface of the type illustrated in Figure 4.11. (Compute the genus and number of boundary components, and then apply Theorem 6.)

Figure 4.11

3 Seifert Surfaces and the Genus of a Knot   The main theorem of this section states that every knot is the boundary of an orientable surface. Consequently, geometric methods can be applied to the general study of knots and not just to particular examples.

☐ **THEOREM 7.** *Every knot is the boundary of an orientable surface.*

PROOF

The proof consists of an explicit construction first described by Seifert. An orientable surface with a given knot as its boundary is now called a *Seifert surface* for the knot.

    The construction begins by fixing an oriented diagram for the knot. Beginning at an arbitrary point on an arc, trace around the diagram in the direction of the orientation. Any time a crossing is met, change arcs along which you trace, but do so in such a way that the tracing continues in the direction of the knot. If at some point you start retracing your path, go to an untraced portion of the diagram and begin tracing again. Figure 4.12 illustrates the result of this procedure for a particular knot.

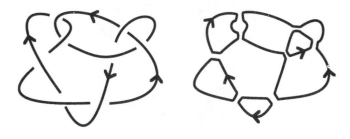

Figure 4.12

The result of this procedure is a collection of circles, called *Seifert circles*, drawn over the diagram. These circles can now be used to construct an orientable surface, as follows.

Each of the circles is the boundary of a disk lying in the plane. If any of the circles are nested, lift the inner disks above outer disks, according to the nesting.

To form the Seifert surface connect the disks together by attaching twisted bands at the points corresponding to crossing points in the original diagram. These bands should be twisted to correspond to the direction of the crossing in the knot. Figure 4.13 illustrates the final surface if this algorithm is applied to the knot in Figure 4.12.

It should be clear that the resulting surface has the original knot as its boundary, that it is orientable is not hard to prove either. (See Exercise 3.3.) Many different surfaces can have the same knot as boundary; stated differently, a knot can have many Seifert surfaces. □

□ **DEFINITION.** *The genus of a knot is the minimum possible genus of a Seifert surface for the knot.*

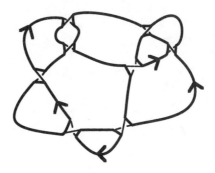

Figure 4.13

For example, Figure 4.1 shows that the trefoil bounds a surface of genus 1. On the other hand, it cannot bound a surface of genus 0, that is a disk, because then it would be unknotted, which is not the case.

A warning is called for here. It can be quite difficult to compute the genus of a knot. The genus of the surface produced by Seifert's algorithm depends on the diagram used, and, more importantly, Seifert's algorithm will not always yield the minimum genus surface! Even with this difficulty the genus is a powerful tool for studying knots.

EXERCISES

3.1. The knot in Figure 4.1 bounds a surface of genus 1, as drawn. What genus surface results if Seifert's algorithm is used to construct a Seifert surface starting with the diagram of the knot given in Figure 4.1?

3.2. Does the surface constructed by Seifert's algorithm depend on the choice of orientation of the knot? What if the procedure was used on a link instead?

3.3. Why does Seifert's algorithm always produce an orientable surface?

3.4. In applying Seifert's algorithm, a collection of Seifert circles is drawn. Express the genus of the resulting surface in terms of the number of these Seifert circles and the number of crossings in the knot diagram.

3.5. A double of a knot K is constructed by replacing K with the curve illustrated in Figure 4.14a. Figure 4.14b illustrates a double of the trefoil knot. The number of twists between the two parallel strands is arbitrary. Show that doubled knots have genus at most 1.

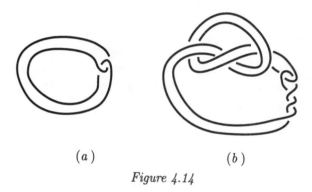

(a) (b)

Figure 4.14

4 Surgery on Surfaces As discussed before,, Seifert surfaces can be very complicated. This section presents *surgery*, a method for simplifying surfaces. All the surfaces that occur later are orientable, and only that case will be described.

Underlying the constructions that follow are two observations. The first is that if two surfaces intersect along intervals, or circles, contained in their boundaries, then the union of the surfaces is again a surface. In the previous section the effect of such constructions on the Euler characteristic and genus was studied. Secondly, note that if one surface is contained in another, and the boundaries are disjoint, then removing the interior of the smaller from the other surface results in a new surface. For example, removing a disk from the interior of a surface results in a surface with one more boundary component. (This construction is sometimes called puncturing the surface.)

SURGERY

The process of cutting out pieces of a surface and pasting on other surfaces forms the basic operation of surgery. The initial set-up is the following. F is a surface in 3-space and D is a disk in 3-space. The interior of D is disjoint from F and the boundary of D lies in the interior F. This is all illustrated in Figure 4.15.

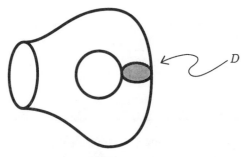

Figure 4.15

The construction of a new surface proceeds as follows. Remove a strip, or annulus, on F along the circle where

F and D meet. The new surface has two more boundary components than F. To each of these boundary components attach a disk which is parallel to the disk D. F has been transformed into a new surface by removing one annulus and adding two disks.

☐ **DEFINITION.** *This procedure is referred to as performing surgery on F along D.*

The effect of surgery on the surface in Figure 4.15 is illustrated below. Note that if the boundary of D had been a different curve on F, then the surface that results from surgery might have had two components. In such cases the curve is called *separating*.

Figure 4.16

What is the effect of surgery on the genus of F? There are two cases to consider. In the first case the new surface has one component. In the second it has two.

☐ **THEOREM 8.** *If surgery on a connected orientable surface, F, results in a connected surface, F', then* genus(F') = genus(F) − 1. *If surgery results in a surface with two components, F' and F'', then* genus(F) = genus(F') + genus(F'').

PROOF

The proof proceeds by computing the effect of the two steps in surgery on the Euler characteristic of the surface. The Euler characteristic of an annulus is 0. Therefore, by Theorem 4, removing the annulus has no effect on the Euler characteristic of the surface.

The Euler characteristic of a disk is 1, so by Theorem 4 the effect of adding on the two disks is to increase the Euler characteristic by 2. Hence, the overall effect of surgery is to increase the Euler characteristic by 2. It follows from the formula for the genus of a connected surface that the genus is then decreased by 1.

In the case that the original surface F is split into two surfaces, F' and F'', the calculation is as follows. Let B, B', and B'' be the number of boundary components of F, F', and F'', respectively. Note that $B = B' + B''$. Hence:

$$
\begin{aligned}
\mathrm{genus}(F') &+ \mathrm{genus}(F'') \\
&= (2 - \chi(F') - B')/2 + (2 - \chi(F'') - B'')/2 \\
&= (4 - \chi(F') - \chi(F'') - B)/2 \\
&= (4 - (\chi(F) + 2) - B)/2 \\
&= \mathrm{genus}(F). \qquad \qquad \square
\end{aligned}
$$

5 Connected Sums of Knots and Prime Decompositions

The connected sum of knots has already appeared in the exercises. It is now time formally to define this construction. The theory of prime knots and the prime decomposition theorem can then be presented.

Suppose that a sphere in 3-space intersects a knot, K, in exactly two points, as illustrated in Figure 4.17. This splits the knot into two arcs. The endpoints of either of those arcs can be joined by an arc lying on the sphere. Two knots, K_1 and K_2, result.

Figure 4.17

☐ **DEFINITION.** *In the situation above K is called the connected sum of K_1 and K_2, denoted $K = K_1 \# K_2$.*

Given two knots, K_1 and K_2, it is easy to construct a knot K such that $K = K_1 \# K_2$. Surprisingly, K is not determined by K_1 and K_2. Examples illustrating the difficulty are hard to construct, but the nature of the problem appears with a discussion of orientation.

If the original knot K is oriented, then both K_1 and K_2 are naturally oriented. Conversely, if K_1 and K_2 are oriented knots it is possible to find a unique oriented knot K such that $K = K_1 \# K_2$ as oriented knots. To come up with a well-defined operation for which the equivalence classes of K_1 and K_2 determines the equivalence class of $K_1 \# K_2$ it is actually necessary to work with oriented knots. For instance, it can be shown that if an oriented knot K is distinct from its reverse, then the oriented connected sum of K with itself is distinct from the oriented connected sum of K with its reverse; that is, distinct even if orientations are ignored.

With connected sum carefully defined, the notion of prime knot can now be introduced, along with the prime decomposition theorem for knots.

□ **DEFINITION.** *A knot is called prime if for any decomposition as a connected sum, one of the factors is unknotted.*

□ **THEOREM 9.** (*Prime Decomposition Theorem*) *Every knot can be decomposed as the connected sum of nontrivial prime knots. If $K = K_1 \# K_2 \# \cdots \# K_n$, and $K = J_1 \# J_2 \# \cdots \# J_m$, with each K_i and J_i nontrivial prime knots, then $m = n$, and, after reordering, each K_i is equivalent to J_i.*

The proof of the existence of a prime decomposition follows immediately from the additivity of knot genus, to be proved below, using induction on the genus of the knot: if a knot decomposes as a nontrivial connected sum, then each factor has lower genus than the original knot; genus 1 knots are prime because 1 is not the sum of positive integers. The uniqueness of decompositions will not be proved here. The complete proof is similar to the proof of additivity of genus, as it involves the careful manipulation of surfaces in 3-space, in particular the families of spheres that split the knot into a connected sum. However, the argument is quite long and detailed.

□ **THEOREM 10.** (*Additivity of knot genus*) *If* $K = K_1 \# K_2$ *then* $genus(K) = genus(K_1) + genus(K_2)$.

PROOF

The proof that the genus of the connected sum is at most the sum of the genera of the summands is easy. Minimal genus Seifert surfaces for K_1 and K_2 can be pieced together to form a Seifert surface for the connected sum. By Corollary 3, the genus of that surface is the sum of the genus of each piece. It remains to show that the surface is in fact a minimal genus Seifert surface for the connected sum.

The argument that the genus of the connected sum is at least the sum of the genera goes as follows. Figure 4.17 illustrates the connected sum of K_1 and K_2 along with a separating sphere S. Let F be a minimal genus Seifert surface for the connected sum. The surface is not drawn as there is initially no information as to how it sits in space relative to K_1, K_2, and S. It will be shown that there is a second surface, G, of the same genus as F, which can be described as the union of Seifert surfaces for K_1 and K_2, meeting in a single interval of their boundaries. It follows from Corollary 3 that the genus of G is the sum of the genera of those two surfaces, and is hence at least the sum of the minimal genera of Seifert surfaces of those knots. The approach is to work with the intersection of F and S. F intersects S in a collection of arcs and circles on S. (Initially, this might not be quite true. For instance, the intersection could contain some isolated points. However, moving F slightly will eliminate any such unexpected intersections.)

In addition, it should be clear that the only arc of intersection on S runs from the two points on S that intersect K. Now one works with the circles of intersection, using surgery to eliminate them one by one.

Consider an innermost circle of intersection. That is, pick one of the circles on S that bounds a disk on S containing no points of intersection of F and S in its interior. Surgery can be performed on F along this disk to construct a new surface bounded by K. If the new surface is connected, then it is a Seifert surface for K, which, by Theorem 8, has lower genus than did F, contradicting the minimality assumption on the genus of F. Hence, surgery results in a disconnected surface. Remove the component that does not contain K. The remaining surface has genus less than or equal to that of F (Theorem 8 again), and by the minimality assumption it actually has the same genus as F. In addition, this new surface will have fewer circles of intersection with S; the circle along which the surgery was done is no longer on the surface.

Repeating this construction, a surface G results that meets S only in an arc. Hence G is formed as the union of Seifert surfaces for K_1 and K_2 that intersect in a single arc, as desired.

This argument is often referred to as a *cut-and-paste* argument, because it consists of cutting out portions of the surface and pasting in new pieces of the surface. Another name for this type of geometric construction is an *innermost circle* argument. This type of argument is typical of geometric proofs in knot theory, and in geometric topology.   □

As described earlier, the existence of prime decompositions follows from the additivity of knot genus; as a knot is decomposed as a connected sum, the genus of the factors decreases. The uniqueness follows from a much more careful cut-and-paste, innermost circle proof. The additivity of genus has the following immediate consequence.

□ **COROLLARY 11.** *If K is nontrivial, there does not exist a knot J such that $K \# J$ is trivial.*

EXERCISES

5.1. Give a proof of the final corollary.

5.2. Use the connected sum of 3 distinct knots to find an example of a knot which can be decomposed as a connected sum in two different ways.

5.3. Prove that a genus n knot is the connected sum of at most n nontrivial knots.

5.4. Fill in the details of the proof of the existence of prime decompositions using the additivity of genus.

5.5. Use the genus to give a simple proof that there are an infinite number of distinct knots. As a much harder problem, can you find an infinite number of distinct prime knots? (Later, once more efficient means are developed to compute Alexander polynomials, this too will become a simple exercise.)

CHAPTER 5
ALGEBRAIC TECHNIQUES

The field of mathematics called algebraic topology is devoted to developing and exploring connections between topology and algebra. In knot theory, the most important connection results from a construction which assigns to each knot a group, called the *fundamental group of the knot*. Knot groups will be developed here using combinatorial methods. An overview of the general definition of the fundamental group is given in the final section of the chapter.

The fundamental group of a nontrivial knot typically is extremely complicated. Fortunately, its properties can be revealed by mapping it onto simpler, finite, groups. The symmetric groups are among the most useful finite groups for this purpose. This chapter begins with a review of symmetric groups. Following that, it is shown how a symmetric group can provide new means of studying knots. The rest of the chapter is devoted to studying the connection between groups and knots more closely.

1 Symmetric Groups   The discussion of symmetric groups that follows focuses on a particular example, S_5. The reader will have no trouble generalizing to S_n, and is asked to do so in the exercises.

Several results that will be used later are described in the exercises also.

Let T denote the set of positive integers, $\{1,2,3,4,5\}$. Recall that a *permutation* of T is simply a one-to-one function from T to itself. There are $5! = 120$ such permutations.

The set of all such permutations, denoted S_5, has an operation defined on it via compositions of functions; the composition of two permutations, thought of as functions, defines a new permutation. The notation for g composed with f is fg. (That is, $fg(i) = g(f(i))$.) This order is reversed from what is often used in algebra, but is fairly standard in knot theory.

As a specific example, suppose that f is the function that sends 1 to 2, 2 to 3, 3 to 4, 4 to 5, and 5 to 1. Let g denote the function that sends 1 to 3, 2 to 4, 3 to 2, 4 to 1, and 5 to 5. Then fg sends 1 to 4, 2 to 2, 3 to 1, 4 to 5, and 5 to 3.

The properties of composition are especially interesting. For instance, note from the start that it is not commutative. In the example above, fg is different from gf. Check this. (As a quick exercise, why is the product associative?)

CYCLIC NOTATION
There is a clever shorthand notation that greatly simplifies working with permutations. It is called *cyclic notation*. A cycle consists of an ordered sequence of distinct elements from T, and represents the permutation that carries each element to the next on the list, sending the last to the first. All the elements that do not appear are fixed by the permutation.

EXAMPLE 1
The symbol $(1,3,4,2,5)$ denotes the permutation that takes

1 to 3, 3 to 4, 4 to 2, 2 to 5, and 5 to 1. It is called a 5-cycle. Note that it represents the same permutation as does the cycle $(3,4,2,5,1)$.

EXAMPLE 2
The symbol $(2,4,5)$ denotes the permutation that takes 2 to 4, 4 to 5, and 5 to 2. It is called a 3-cycle. The terms that do not appear are fixed by the corresponding permutation. That is, 1 goes to 1 and 3 goes to 3.

The following is an especially useful theorem.

□ **THEOREM 1.** *Every permutation can be written as the product of cycles, no two of which have an element in common.*

The proof of this is given in most introductory texts in algebra. The exercises give some practice in writing permutations as such products, and with a little work the notation will become second nature.

EXAMPLE 3
The permutation that takes 1 to 3, 3 to 2, 2 to 1, 4 to 5, and 5 to 4 can be written as $(1,3,2)(4,5)$, the product of a disjoint 3-cycle and a 2-cycle.

Using cyclic notation it is also easy to write down and compute the product of permutations.

EXAMPLE 4
$(1,3,2)(2,3)(1,5,4) = (1,2,5,4)$. (For instance, since the first cycle sends 1 to 3 and the second sends 3 to 2, and the last does not affect 2, the composition sends 1 to 2. The first sends 3 to 2 and the second sends 2 to 3, and the last does not affect 3, so the composition sends 3 to 3.)

EXAMPLE 5
This is really one more exercise.
$(1,3,4)(1,4,5)(2,3)(1,3,2,5,4)(1,4)(2,5,3) = (1,3)(2,5)$.

GENERATING SUBSETS
A set of permutations, $\{g_1,\ldots,g_k\}$ is said to *generate* the symmetric group if every element in the group can be written as a product of elements from the set, with possible repetitions, and their inverses. In Exercise 7 it is shown that certain sets of *transpositions* (i.e., 2-cycles) generate the symmetric group. Exercise 9 presents other generating sets.

NOTATION
This cyclic notation varies from reference to reference. First, consider the permutation that sends 1 to 3, 3 to 1, 2 to 2, 4 to 4, and 5 to 5. It is written here as $(1,3)$. Some books write it as $(1,3)(2)(4)(5)$. This added notation is useful in indicating that the original set T contained the elements $\{1,2,3,4,5\}$.

Second, note again that in the notation used here, permutations are multiplied from left to right. In many references they are multiplied from right to left.

Finally, there is some ambiguity in the notation. Does the symbol $(1,2,3)(4,5)$ denote a single permutation, or the product of two cycles? In either case, the actual permutation that is represented is the same. Hence, what appears as an ambiguity in notation is actually clear in meaning. Anytime two permutations are written side by side they will be viewed as representing a product.

For the knot theory that follows, facility with the symmetric groups and cyclic notation is essential. The following exercises provide practice in working with the symmetric groups and also describe important results that will be used in the next sections.

EXERCISES

The definition of the symmetric group S_6, or for that matter S_n, should now be clear; it exactly corresponds to everything above, with 5 replaced by 6, or by n.

1.1. This first exercise concerns some explicit calculations in the group S_6, the set of permutations of the set $\{1,2,3,4,5,6\}$.

(a) Let f be the permutation given by $f(1) = 4$, $f(2) = 3$, $f(3) = 6$, $f(4) = 5$, $f(5) = 2$, $f(6) = 1$. Write f in cyclic notation.

(b) Same question for g, where g is given by $g(1) = 5$, $g(2) = 1$, $g(3) = 3$, $g(4) = 6$, $g(5) = 2$, $g(6) = 4$.

(c) Simplify the following products. That is, write each as a product of disjoint cycles.

    i. $(1,2,3)(4,5,6)(1,2)(3,4)(5,6)$,

    ii. $(1,2)(3,5,6,4,)(1,3,5)(4,2)$,

    iii. $(1,2)(3,6,4,5)(1,3,6)(1,3,5)(2,4)$,

    iv. $(1,2)(2,3)(3,4)(4,5)(5,6)$.

1.2. Show that S_6 is not commutative. That is, find permutations f and g such that fg does not equal gf. (Write f and g in cyclic notation.)

1.3. Again working in S_6,

(a) The inverse to $(1,4,2,5)(3,6)$ is $(1,5,2,4)(3,6)$. Show this. (That is, verify that the product of these two permutations is the *identity* permutation. The identity permutation is the permutation f that satisfies $f(x) = x$ for all x in the domain.)

(b) Find the inverses to $(1,3,6,4,5,2)$, $(1,6,4)(2,5,3)$, and $(1,2,3,4)(3,4,2)(3,5,6,1)$.

(c) In general, how does one write down the inverse of a permutation given in cyclic notation?

(d) Let g be the permutation $(1,6,3)(2,4,5)$. Compute $g^{-1}(2,3,4)g$, and $g^{-1}(1,3)(4,5,2,6)g$. What is a short cut for computing $g^{-1}fg$ in general? (Hint: It might help to recall that in cyclic notation different looking permutations can be the same. For instance, $(1,5,3,4)$ $= (5,3,4,1) = (3,4,1,5) = (4,1,5,3)$.) The operation of going from f to $g^{-1}fg$ is called *conjugation* by g.

1.4. Is it possible in S_8 for the product of two 4-cycles to be an 8-cycle?

1.5. The *order* of a permutation f is the least positive integer n such that f composed with itself n times is the identity.

(a) What is the order of the cycle $(1,3,4,6,2)$?

(b) Verify that the order of $(1,3,5)(2,4)$ is 6.

(c) What is the largest order of an element in S_7? In S_{10}? In S_{20}?

1.6. Find all of the 4-cycles in S_4 that commute with the cycle $(1,2,3,4)$. Describe the cycle structure of the permutations of S_8 which commute with $(1,2,3,4)$.

1.7. (a) Check that the 5-cycle $(1,2,3,4,5)$ is equal to the product of transpositions, $(1,2)(1,3)(1,4)(1,5)$.

(b) Write $(1,5,3,4)$ as a product of four transpositions. Write $(1,2,4)(3,5,6)$ as a product of transpositions.

(c) Argue that every permutation is the product of transpositions, and more specifically is a product of transpositions of the form $(1,n)$.

(d) Show that every permutation can be written as the product of transpositions taken from the set
$\{(1,2),(2,3),(3,4),(4,5),\ldots,(n-1,n)\}$.

1.8. What is the least number of transpositions that can generate S_n?

1.9. (a) Show that the set of 4-cycles generates S_4. (Find a way to write a single transposition as a product of 4-cycles.)

(b) Show that the 4 cycles $(1,2,3,4)$ and $(1,2,4,3)$ generate S_4.

1.10. A permutation can be written as the product of transpositions in many ways. A basic result about the symmetric groups states that the parity of the number of transpositions used does not depend on the choice of expansion. For instance, since $(1,3,5)(2,4,6,7) = (1,3)(1,5)(2,4)(2,6)(2,7)$, any expansion of $(1,3,5)(2,4,6,7)$ will use an odd number of transpositions. The *sign of a permutation* is said to be even or odd depending on whether or not an even or odd number of transpositions appears in its expansion.

(a) Show that the sign of an n-cycle is even if and only if n is odd.

(b) How does the sign of a product of permutations depend on that of its factors?

2 Knots and Groups In Chapter 3 knot diagrams were labeled in a variety of ways. Now a procedure will be described for labeling knots with elements of a group. The discussion could be simplified by using some specific group. For example, it may be helpful to think of G as S_5 or S_n on the first reading of this section. The examples and exercises will mostly be taken from S_n.

It was seen in Chapter 3 that labelings mod p provide a powerful means for studying knots. Each odd prime number offered a potential tool for studying a knot. Now it will be seen that each group offers another potential tool. The subtle and intricate properties of a group can reflect the details of complicated knotting.

The work in this section will be done with oriented knots. As in the case of the Alexander polynomial the results do not depend on the choice of orientation. However in this case the independence on orientation is easily proved.

LABELING KNOT DIAGRAMS
A labeling of an oriented knot diagram with elements of a group consists of assigning an element of the group to each arc of the diagram, subject to the following two conditions.

(1) *Consistency*: At each crossing of the diagram three arcs appear, each of which should be labeled with an element from the group. Suppose the labels are the group elements g, h, and k. In the case of a right-handed crossing as illustrated in Figure 5.1a, the labels must satisfy $gkg^{-1} = h$.

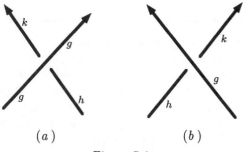

(a) (b)

Figure 5.1

At a left-handed crossing, as illustrated in Figure 5.1b, the condition is that $ghg^{-1} = k$.

(2) *Generation*: The labels must generate the group. (As described in the previous section, this means that every element in the group can be written as a product of the elements that appear as labels, along with their inverses.)

Given a set of elements in a group, it is often difficult to decide if they form a generating set. The exercises in the previous section gave some examples of sets of permutations that generate S_n. More examples will follow. If the notion of generation is not yet clear, focus on the consistency condition. With some practice the idea of generation will become clear as well.

Figure 5.2 indicates labelings of the edges of the trefoil knot with elements from the symmetric group S_3 and from S_4. It is straightforward to check that at each crossing the consistency condition is satisfied. All the transpositions of S_3 appear, so the set of labels does satisfy the generation condition also. Exercise 9 of the previous section shows that the second set of labels also generates.

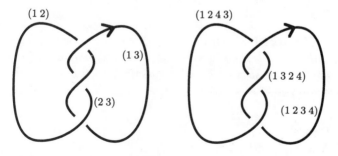

Figure 5.2

As a second example consider the knot in Figure 5.3. In that diagram a labeling from S_5 is indicated. Again the two conditions can be verified.

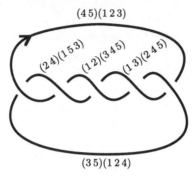

Figure 5.3

The usefulness of these labelings comes from the following theorem. It states that if some diagram of a knot can be labeled with elements of a group G, then every diagram for that knot can be labeled with elements of G. Hence, every example of a group provides a potentially new means for distinguishing knots. For instance, the knot in Figure 5.3 is nontrivial since it can be labeled with elements from S_5 while the unknot cannot be. To state the theorem formally:

☐ **THEOREM 1.** *If a diagram for a knot can be labeled with elements from a group G, then any diagram of the knot can be so labeled with elements from that group, regardless of the choice of orientation.*

PROOF
A combinatorial proof of this theorem is available. As in the proofs of Chapter 2, one just checks what happens with

each Reidemeister move. The proof has a large number of cases, none of which are difficult.

Figure 5.4 indicates a portion of a knot diagram before and after a Reidemeister move has been performed. Labelings on each diagram indicate how the labeling on the first diagram can be changed into labelings on the second. You are invited to check a few more cases of this combinatorial proof. Recall that there are cases corresponding to other Reidemeister moves and also cases corresponding to the same moves but with different orientations on the edges.

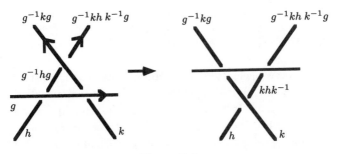

Figure 5.4

Checking that the choice of orientation does not matter is fairly easy. If an oriented diagram of a knot can be labeled with elements from a group G, the same diagram of the knot, with its orientation reversed, can be labeled with elements of G by just labeling each edge with the inverse of the element that was used in the first diagram. The result follows from observations of the type that if $gkg^{-1} = h$ then $g^{-1}h^{-1}g = k^{-1}$. $\qquad\square$

The use of labelings is one of the most powerful means of distinguishing knots. For instance, in Thistlethwaite's compilation of prime knots with 13 crossings, he found 12,965 knots. However, only 5,639 different Alexander polynomials appear and 14 have polynomial 1. By using labelings (taken from thirteen different groups) he was able to reduce the number of unidentified knots down to about 1,000. A more refined approach, based on results to be described in the next section, gave Thistlethwaite's complete classification.

Although the definition of a labeling with group elements is relatively simple, actually finding such labelings can be extremely difficult. There is one important observation that simplifies the process. At a crossing in a diagram, once the overcrossing and one of the two other arcs are labeled, the label on the last arc is forced by the consistency condition.

Many fascinating results and problems in knot theory concern labelings with group elements. For instance, Perko proved a remarkable theorem stating that, if a knot can be labeled with elements from S_3, it also has an S_4 labeling. Similar results concerning other groups have since been discovered.

EXERCISES

2.1. Check that the consistency condition is satisfied at all the crossings in the labeled knot diagrams illustrated by Figures 5.2 and 5.3.

2.2. In Figure 5.5 two of the labelings satisfy the consistency condition, while one of the three does not. Find the inconsistent labeling.

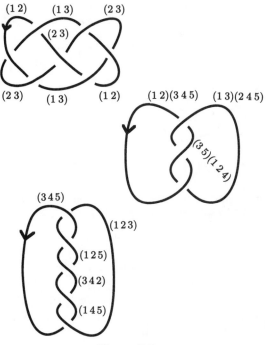

Figure 5.5

2.3. Figure 5.6 illustrates a knot with some of its edges labeled. Use the consistency condition to determine a labeling for the entire diagram.

2.4. Find a labeling of the $(3,3,3)$-pretzel knot illustrated in Figure 5.7 using transpositions from S_4. Your labeling should have every transposition appear, so it is clear that the labels generate the group. (Hint: Pick your labels for the three strands indicated. This will force a choice of the rest of the labels via the consistency condition.)

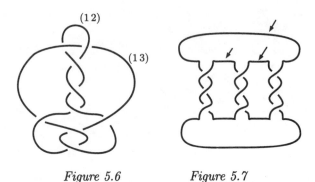

Figure 5.6   *Figure 5.7*

2.5. For what values of p, q, and r, can the (p,q,r)-pretzel knot be labeled with transpositions from S_4?

2.6. In order to make sure that the independence of orientation is clear, do the following exercise. Check that in some of the previous examples the labeling becomes inconsistent if the orientation of the knot is reversed. Show, however, that if the orientation of the knot is reversed and if each label is replaced with its inverse, then the labeling will again become consistent.

2.7. Suppose that a knot diagram is labeled with elements in a group, and g is some arbitrary element in the group. Show that if each label, ℓ, is replaced with its conjugate by g, $(g^{-1}\ell g)$, then the resulting labeling satisfies the consistency condition.

3 Conjugation and the Labeling Theorem   If a knot is labeled with elements from a group, all the labels that appear represent conjugate elements of the group. This simple observation can be used to add great power to the method of labeling. Thistlethwaite's classification of 13 crossing knots depended heavily on the inclusion of conjugacy considerations. More important, this added detail provided the first means of showing that a particular oriented knot is not equivalent to its reverse.

CONJUGACY RELATIONS IN A GROUP

Elements g and h in a group G are called conjugate if there is an element k in G such that $k^{-1} \cdot g \cdot k = h$. For example, in S_5 the element $(1,2)(3,4,5)$ is conjugate to $(2,4)(1,5,3)$. Just let $k = (1,2,4,5,3)$.

In the symmetric group, two elements are conjugate if and only if they have the same *cyclic structure*. That is, a product of a 2-cycle and a disjoint 3-cycle is conjugate to any other such product, but is never conjugate to a 5-cycle or the product of two disjoint 2-cycles. This follows from the reasoning used in Exercise 1.3 of this chapter.

Using the notion of conjugacy, a group can be broken down into conjugacy classes consisting of all conjugate elements in the group. In S_5 there are 7 conjugacy classes: elements conjugate to $(1,2)$; $(1,2,3)$; $(1,2,3,4)$; $(1,2,3,4,5)$; $(1,2)(3,4)$; $(1,2)(3,4,5)$; and elements conjugate to the identity element.

CONJUGACY AND LABELINGS

Suppose a knot diagram is labeled with the elements of a group. At each crossing the consistency conditions imply that the label of the arc that passes under the crossing is conjugate to the one that emerges from the crossing. It follows that all the labels of the labeling are conjugate elements of the group. As an example, the knot diagram for 9_{46} in the appendix can be labeled with transpositions from S_4 while such a labeling is not possible for the diagram of 6_1. On the other hand, the diagram for 6_1 can be labeled with 4-cycles from S_4 while no such labeling is possible for the diagram of 9_{46}. (See Exercise 3.4.)

The main result of the previous section can now be strengthened. The proof of the following theorem is similar to that of the proof of Theorem 1. Note however that orientation is now an issue. There are examples of groups for which not every element is conjugate to its inverse. These groups provide one of the few means of distinguishing a knot from its reverse.

□ **THEOREM 2.** *If a diagram of an oriented knot can be labeled with elements of a group, with the labels coming from some conjugacy class of the group, then every diagram of that knot can be labeled with elements from that conjugacy class.*

To see the usefulness of this theorem to the problem of classifying knots consider the previous example of the knots 6_1 and 9_{46}. The fact that one can be labeled using transpositions from S_4 and the other cannot proves that the knots are distinct. This is an especially interesting example since these knots cannot be distinguished using colorings, and both have the same Alexander polynomial, $-2t^2 + 5t - 2$.

It is often the case that if such tools as the polynomial and geometric techniques cannot distinguish two knots, then some clever choice of groups and labelings will do the trick.

EXERCISES

3.1. How many conjugacy classes are there in S_6?

3.2. Prove that it is impossible to find a labeling of the trefoil using transpositions from S_4. (Check that once the labels on two edges are determined, all the labels are determined, and then apply the result of Exercise 1.8.)

3.3. (a) Show that if an oriented knot is equivalent to its reverse and can be labeled so that some edge is labeled with a group element g, then it also has a labeling with some label g^{-1}.

(b) Show that in the symmetric group every element is conjugate to its inverse. Hence, the labeling theorem applied using the symmetric group alone is not sufficient to distinguish a knot from its reverse.

(c) The set of even permutations forms a subgroup of S_n, called the alternating group, and denoted A_n. Show that a 7-cycle is not conjugate to its inverse in A_7. (To conjugate a 7-cycle to its inverse in S_7 requires an odd permutation to do the conjugating.)

3.4. Check the claims about labelings of the diagrams for 6_1 and 9_{46}.

3.5. The theory of labelings with group elements applies to links as well as to knots. (Why?) It is not true, however, that all the labels now come from the same conjugacy class. This is easily demonstrated with the unlink. Prove that the labels on each component of a labeled link are conjugate.

---

4 Equations in Groups and the Group of a Knot   Finding labelings of a knot using elements of a group is apparently quite difficult, especially if you proceed by guesswork. In the exercises it was seen that by taking advantage of the consistency relationships, once a few labels are chosen, the rest are determined. This approach can be formalized to reduce the problem of finding labelings into one of solving equations in a group. In addition to providing a practical tool for studying knots, the equations can be used to actually define the fundamental group of the knot.

Consider the knot in Figure 5.8. Fix a group G to be used in the labeling. (For now you might want to think of some particular symmetric group.) Suppose that labels x, y, and z are picked for the top three arcs. Using the consistency condition, it follows that the next arcs must be labeled as indicated in the figure. From there you can proceed down the knot. Moving down, each crossing determines a label on another arc. The labeling of each arc is forced by the labels that preceded it. Finally, using the marked crossings you end up with the labeling indicated in Figure 5.8.

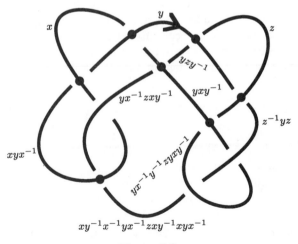

Figure 5.8

Clearly, the procedure has produced a consistent labeling if the consistency condition holds at the remaining three crossings. That is, the knot has been consistently labeled if the equations;

$$xyx^{-1} = yx^{-1}zxy^{-1}xyx^{-1}z^{-1}xy^{-1}$$

$$xy^{-1}x^{-1}yx^{-1}zxy^{-1}xyx^{-1}$$
$$= yx^{-1}y^{-1}zyxy^{-1}z^{-1}yzyx^{-1}y^{-1}z^{-1}yxy^{-1}$$

$$yx^{-1}y^{-1}zyxy^{-1} = z^{-1}y^{-1}zyxy^{-1}z^{-1}yz$$

hold in G.

The labels of the diagram will generate the group if and only if x, y, and z generate the group. (Why?) In summary, the knot pictured can be labeled with elements from G if and only if there are generators for G, x, y, and z, satisfying the equations

$$yx^{-1}zxy^{-1}xyx^{-1}z^{-1}xy^{-1}xy^{-1}x^{-1} = 1$$

$$yx^{-1}y^{-1}zyxy^{-1}z^{-1}yzyx^{-1}y^{-1}z^{-1}yxy^{-1}$$
$$xy^{-1}x^{-1}yx^{-1}z^{-1}xy^{-1}xyx^{-1} = 1$$

$$z^{-1}y^{-1}zyxy^{-1}z^{-1}yzyx^{-1}y^{-1}z^{-1}yxy^{-1} = 1$$

(See Exercise 4.1 concerning the equivalence of the two sets of equations.)

In general, finding a labeling for a knot can always be reduced to solving equations in the group. This could have been pointed out as soon as labeling was defined since finding a labeling for a knot with an n crossing diagram is equivalent to solving n equations (arising from the consistency condition at each crossing) in n variables (the labels on the arcs of the diagram.) However, the procedure just

given usually results in many fewer, though more complicated, equations.

(The reduction in the number of equations could be
carried out algebraically. Some of the original n equations
express individual variables in terms of others. Those variables could then be removed from the system of equations
using substitutions. The approach just given is usually
simpler.)

It is worth noting that the three equations that arose
in the preceding example are redundant. That is, if two
hold the other one is automatically true. (Can you see
why?) This is generally the case. Initially the number of
variables and equations will be equal, but it turns out that
any one of the equations is a consequence of the others.

PRESENTATIONS OF GROUPS, THE GROUP OF A KNOT

A detailed description of presentations of groups lies in the
realm of combinatorial group theory. The basic construction is easily summarized. Any collection of variables along
with a set of *words* in those variables defines a group. A
word is just an expression formed from the variables and
their inverses. The set of variables and words is said to
give a *presentation* of the group; the variables are called
the *generators* of the group, and the equations formed by
setting the words equal to 1 are called the defining *relations* of the group. Informally, the group is defined as
follows. An element consists of a word in the variables
and their inverses. Multiplication of such words is carried
out by concatenation, that is putting one word after the
other. The identity element is given by the "empty" word,
and is usually denoted "1". Finally, two words are considered equivalent if one can be obtained from the other
by repeatedly (1) either adding or removing variables fol-
lowed by their inverses, and (2) either adding or deleting
appearances of the defining words.

EXAMPLE

The two variables x and y, along with the word $xyxy^{-1}x^{-1}y^{-1}$ give the presentation of a group. It is written as $G = \langle x,y \mid xyxy^{-1}x^{-1}y^{-1} = 1 \rangle$. Let $g = xy$, and $h = yxy$. Then in G the relation $g^3 = h^2$ holds. To see this, write

$$g^3 = (xy)(xy)(xy) = xyxyxy = xyx(y^{-1}y)yxy$$
$$= xyxy^{-1}yyxy = xyxy^{-1}(x^{-1}x)yyxy$$
$$= xyxy^{-1}x^{-1}xyyxy$$
$$= xyxy^{-1}x^{-1}(y^{-1}y)xyyxy$$
$$= (xyxy^{-1}x^{-1}y^{-1})yxyyxy = yxyyxy = h^2.$$

One remark about this example. A second group could be defined by the presentation $\langle g, h \mid g^3h^{-2} = 1 \rangle$. The calculation just given essentially proves that the two groups are isomorphic.

There are many shortcuts available for working with groups and their presentations, and the calculations above could be simplified. The exercises on this material are not essential for continuing. In doing the exercises the reader will discover some of the shortcuts and will also develop intuition about group presentations.

THE GROUP OF KNOT

Given a knot diagram we have seen that it is possible to come up with a collection of variables and equations of the form $w_i = 1$. Furthermore, given such a set of variables and equations a group naturally arises. This group is called the *group of the knot*. Although the group itself depends on the choices made, such as the choice of the diagram, it can be proved that any two groups that arise in this way for a given knot will be isomorphic.

The study of knot groups is a central topic in knot theory. One of the most significant results in the subject, Dehn's Lemma, was proved by Papakyriakopoulos. It states that if a knot group is isomorphic to Z, the group of integers, then the knot is trivial. Examples of distinct knots with the same group do occur, but it is now known that this is impossible for prime knots. That is, the only knots which are not determined by their knot groups are connected sums of nontrivial knots.

EXERCISES

4.1. Explain why the relation

$$xyx^{-1} = yx^{-1}zxy^{-1}xyx^{-1}z^{-1}xy^{-1}$$

is equivalent to the relation

$$z^{-1}y^{-1}zyxy^{-1}z^{-1}yzyx^{-1}y^{-1}z^{-1}yxy^{-1} = 1.$$

(In general, the relation $g = h$ is equivalent to $gh^{-1} = 1$.)

4.2. In Figure 5.9 two knot diagrams are shown, along with a labeling of some of the edges. Compute the remaining

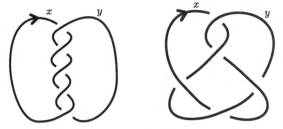

Figure 5.9

labels. In each case the knot group has two generators, or variables, and is determined by a single relation, given along with each diagram. Check these.

5 The Fundamental Group

For any space X, the "fundamental group" is a group that is naturally associated to the space. In studying knots, the space of interest is the complement of the knot in three space, $R^3 - K$. It is not possible to develop the theory in detail here, but the definitions can be summarized. Up until now, the association of algebraic invariants to a knot has depended on the use of the knot diagram, although in each case using the Reidemeister moves it is possible to prove that the final result does not depend on the choice of diagram. With the use of the fundamental group it is possible to define these algebraic quantities, and in particular the group of the knot, without reference to diagrams. There are practical advantages to this approach. For one, it permits the algebraic methods to be applied in settings other than knots in three space. Chapter 9 discusses knots in higher dimensions. Second, it brings to bear many of powerful techniques of algebraic topology, for instance, covering spaces and homology theory. The material presented in this section is not used in the rest of the text.

Denote the complement of a knot K by X, and fix a point p in X. The elements of the fundamental group of X are equivalence classes of closed oriented paths in X which begin and end at p. These paths need not be simple; they may have self-intersections. Two such paths are viewed as equivalent if one can be continuously moved into the other, at all times keeping the endpoints at p. This transformation is called a homotopy. Unlike deformations of knots, in a homotopy the path may have self-intersections. However, at no time may the path leave X; that is, it may not cross K. In Figure 5.10 three paths in the complement of the trefoil path are shown. The paths γ_1 and γ_2 are homotopic, but γ_3 is not homotopic to either one; this may look clear, but is not easy to prove.

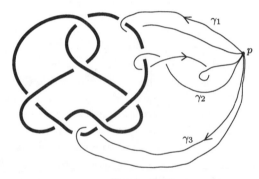

Figure 5.10

These equivalence classes of paths form the elements of the fundamental group; the product of two such elements must now be defined. Given paths γ_1 and γ_2, one can form a new path which travels around γ_1 and then around γ_2. In defining this formally, parametrizations must be discussed. As one example of a product consider the curves γ_1 and γ_3 in Figure 5.10. Their product is homotopic to the path shown in Figure 5.11.

There is a good deal of work involved in proving that the group just described is well defined. Many of the de-

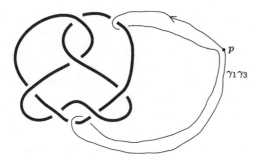

Figure 5.11

tails are concerned with the issue of parametrization. One example of something that must be proved is that if paths γ_1 and γ_2 are homotopic to ω_1 and ω_2, respectively, then the products are homotopic also. Another important part of the proof is the construction of inverses. (The identity element is represented by the constant path at p.)

The definition of the fundamental group is quite abstract, and not very useful for doing calculations. A variety of theorems permit simplifications in its calculation. The most important of these is called the Van Kampen Theorem; it describes how a decomposition of a space leads to a decomposition of the fundamental group.

THE KNOT GROUP, THE FUNDAMENTAL GROUP, AND LABELINGS

Here is a quick summary of the connections between the fundamental group and the algebra presented earlier in the chapter. A diagram of a knot yields a decomposition of the knot complement which, using the Van Kampen Theorem, in turn produces a simple presentation of the fundamental group. That presentation is the same as the presentation of the knot group described in the previous section.

The connection with labeling can also be summarized. For each arc in the diagram of the knot there is an element in the fundamental group which is represented by a path that runs from the base point, p, directly to the arc, once around the arc, and then back to the basepoint. That element corresponds to the element in the knot group given by the variable label on the arc. It can be proved that relations between the elements in the fundamental group correspond to the relations in the knot group arising at the crossings.

A group is often studied by mapping it homomorphically onto a simpler group, say G, which is better understood. Given such a homomorphism of the fundamental group of a knot complement, composing it with the correspondence between the knot group and the fundamental group gives an assignment of an element in G to each arc in the diagram. That is, labelings of the diagram turn out to correspond to homomorphisms of the fundamental group of the knot complement. The consistency condition on the labeling corresponds to the map being a homomorphism. The generation condition corresponds to the map being surjective.

CHAPTER 6:
GEOMETRY, ALGEBRA, AND
THE ALEXANDER POLYNOMIAL

The discovery of connections between the various techniques of knot theory is one of the recurring themes in this subject. These relationships can be surprising, and have led to many new insights and developments. A recent example of this occurred with the discovery by V. Jones of a new polynomial invariant of knots . Although his approach was algebraic, the Jones polynomial was soon reinterpreted combinatorially. Almost immediately there blossomed an array of new combinatorial knot invariants which appear to be among the most useful tools available for problems relating to the classification of knots. An understanding of these new invariants from a noncombinatorial perspective is now a major problem in the subject, and one that will certainly lead to significant progress. Chapter 10 is devoted to a discussion of the Jones polynomial and its generalizations.

To demonstrate how various techniques can be related, this chapter presents geometric and algebraic approaches to the Alexander polynomial. The geometric approach introduces a new and powerful object, the *Seifert matrix*, and for this reason geometry will be the main focus here. The algebraic approach links the combinatorics to the geometry, and also demonstrates that the Alexander polynomial of a knot is determined by the knot group.

It is not surprising that bringing together the diverse methods developed so far involves difficult technical arguments. Even the definition of the Seifert matrix, given in Section 1, is fairly complicated. The benefit of this technical argument is seen in Section 2, where a simple algorithm for computing the Alexander matrix is given, and in Section 3, where new knot invariants are developed. Fox derivatives and their use in computing the Alexander polynomial from a presentation of the knot group are described in Section 4. This material may also appear quite technical; but again there are valuable insights gained from the approach.

1 The Seifert Matrix   If a surface is formed by adding bands to a disk, the cores of the bands along with arcs on the disk can be used to construct a family of oriented curves on the surface. This is illustrated in Figure 6.1. The choice of orientation of the curves is arbitrary. In the case where the surface is a Seifert surface for a knot, how these curves twist and link carries information about the knot. This linking and twisting information is captured by a matrix called the *Seifert matrix* of the knot.

In Exercise 2.4 of Chapter 3, linking numbers were defined. Exercise 2.5 of that chapter provided an alternative definition that is now summarized. Suppose that an oriented link of two components, K and J, has a regular projection. The *linking number* of K and J is defined to be the sum of the signs of the crossing points in the diagram

at which K crosses over J. The sign of a crossing is 1 if
the crossing is right-handed, that is, if J crosses under K
from the right to the left. The sign is -1 if the crossing is
left-handed. The linking number is denoted $\ell k(K, J)$ and
is symmetric: $\ell k(K, J) = \ell k(J, K)$.

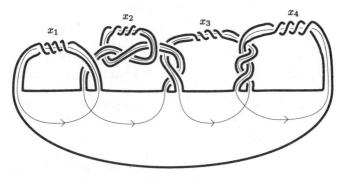

Figure 6.1

Given a knot K, fix a Seifert surface F for K. Since
a Seifert surface is orientable, it is possible to distinguish
one side of the surface as the "top" side. Formally this
consists of picking a nonvanishing normal vector to the
surface. Which direction is picked will not matter. With
this done, given any simple oriented curve, x, on the Seifert
surface, one can form the *positive push off* of x, denoted
x^*, which runs parallel to x and lies just above the Seifert
surface.

If the Seifert surface F is formed from a single disk
by adding bands, it was shown in Figure 6.1 that there
naturally arises a family of curves on F. If F is genus
g there will be $2g$ curves, $x_1, x_2, \ldots, x_{2g}$. The associated
Seifert matrix is the $2g \times 2g$ matrix V with (i, j)-entry $v_{i,j}$
given by $v_{i,j} = \ell k(x_i, x_j^*)$. The Seifert matrix clearly de-

pends on the choices made in its definition, and by itself is not an invariant of the knot. However, in the next two sections it will be shown that the Seifert matrix can be used to define knot invariants, including the Alexander polynomial. The rest of this section is devoted to illustrating the computation of entries in a Seifert matrix.

EXAMPLE

Figure 6.2

Computing the entries of a Seifert matrix can be difficult, especially if the surface is very complicated. Let's consider the Seifert matrix for the Seifert surface and knot illustrated in Figure 6.1. The way the surface is oriented, the normal vector to the surface points toward the reader on the disk portion of the surface. Figure 6.2 illustrates the curves x_2 and x_3^*. Their linking number is 1, so that $v_{2,3} = 1$.

In Figure 6.3 the curves x_2 and x_2^* are drawn. The reader should redraw Figure 6.1 and check that the curve drawn as x_2^* actually lies above the Seifert surface. It is a delicate construction.

Using Figure 6.3, one computes $v_{2,2} = \ell k(x_2, x_2^*) = -5$. Continuing in this way (see Exercise 1.2) the final result is that the Seifert matrix

Figure 6.3

is given by

$$\begin{pmatrix} 2 & 1 & 0 & 0 \\ -5 & -5 & 1 & 0 \\ 0 & 1 & 2 & -1 \\ 0 & 0 & -2 & -2 \end{pmatrix}$$

EXERCISES

1.1. In Figure 6.4 Seifert surfaces for the trefoil knot and its mirror image, the left-handed trefoil, are illustrated. Compute the Seifert matrix associated to each of these surfaces.

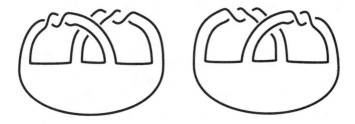

Figure 6.4

1.2. Complete the calculation of the Seifert matrix for the knot in Figure 6.1.

1.3. Figure 6.5 illustrates the Seifert surface of a knot, previously discussed in Exercise 2.2 of Chapter 3. (This particular example is the *3-twisted double* of the unknot.) Compute its Seifert matrix.

1.4. In Exercise 2.5 of Chapter 4 Seifert surfaces for the (p,q,r)-pretzel knot were constructed, for p, q, and r odd. Find the corresponding Seifert matrix.

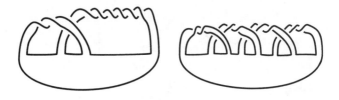

Figure 6.5   *Figure 6.6*

1.5. Figure 6.6 above shows a Seifert surface for the $(2,n)$-torus knot. (Only the $(2,5)$-torus knot is shown, but the pattern is clear.) Find the corresponding Seifert matrix.

1.6. What would be the effect of changing the orientation of the Seifert surface on the Seifert matrix?

1.7. Seifert surfaces for two knots can be used in order to form a Seifert surface for the connected sum of the knots. How are the corresponding Seifert matrices related?

1.8. In Exercise 1, the example of the trefoil and its mirror image can be generalized. What is the relation between the Seifert matrix of a knot, found using some given Seifert surface, and the Seifert matrix for its mirror image, found using the mirror image of the given Seifert surface?

---

2 Seifert Matrices and the Alexander Polynomial

The Alexander polynomial is easily computed using the Seifert matrix; recall, once again, that the polynomial is only defined up to multiples of $\pm t^i$. An immediate con-

sequence will be a proof that the Alexander polynomial
is symmetric. A proof of this based on the combinatorial
definition of the Alexander polynomial is not at all evident.

☐ **THEOREM 1.**   *Let V be a Seifert matrix for a knot
K, and V^t be its transpose. The Alexander polynomial is
given by the determinant, $\det(V - tV^t)$.*

Later in this section it will be indicated why this determinant gives a well-defined knot invariant. The proof
that it is the same as the combinatorially defined Alexander polynomial is a deeper result. The connection is via
algebra: the complement of the knot can be decomposed
using a Seifert surface and that decomposition leads to
information about the structure of the knot group. In Section 4 a connection between the group of the knot and the
Alexander polynomial will be presented. Carefully putting
all these connections together yields the desired result.

One important corollary of Theorem 1 is the following.

☐ **COROLLARY 2.**   *The Alexander polynomial of a knot
K satisfies $A_K(t) = t^{\pm i} A_K(t^{-1})$ for some integer i.*

PROOF
This is an immediate consequence of the fact that a matrix and its transpose have the same determinant: if
a Seifert matrix V is used to compute the Alexander
polynomial $A_K(t) = \det(V - tV^t) = \det((V - tV^t)^t) =
\det(V^t - tV) = \det(tV - V^t) = \det(t(V - t^{-1}V^t)) =
t^{2g}A_K(t^{-1})$.   ☐

S-EQUIVALENCE OF SEIFERT MATRICES
The construction of the Seifert matrix of a knot depended
on many choices. Two of these are especially critical.

Band moves: If a Seifert surface is presented as a disk with bands added, that surface can be deformed by sliding one of the points at which a band is attached over another band. The resulting surface is again a disk with bands added. However, the $2g$ curves formed from the cores of the new bands will not be the same as those formed from the cores of the original bands. The effect of this operation is to do a simultaneous row and column operation on the Seifert matrix; that is, for some i and j, a multiple of the i-th row is added to the j-th row, and then the same multiple of the i-th column is added to the j-th column. A sequence of these band slides changes the Seifert matrix from V to MVM^t where M is some invertible integer matrix.

Stabilization: Given a Seifert surface for a knot, it can be modified by adding two new bands, as illustrated in Figure 6.7 for the Seifert surface of the trefoil. One of the bands is untwisted and unknotted. The other can be twisted, or knotted, and can link the other bands.

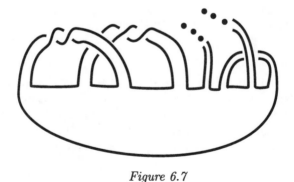

Figure 6.7

It is clear that the boundary of the new surface is the same knot as for the original Seifert surface. The effect of this operation on the Seifert matrix is to add two new columns and rows, with entries as indicated.

$$\begin{pmatrix} & & & & * & 0 \\ & & & & * & 0 \\ & & & & * & 0 \\ & & & & * & 0 \\ * & * & * & * & 1 \\ 0 & 0 & 0 & 0 & 0 \end{pmatrix}$$

Two integer matrices are called S-equivalent if they differ by a sequence of operations of the two types described: right and left multiplication by an invertible integer matrix and its transpose, and addition or removal of a pair of rows or columns of the type shown above. These two matrix operations also include the changes that occur in a Seifert matrix if the bands are reordered, or reoriented.

A difficult geometric argument shows that for any two Seifert surfaces for a knot, there is a sequence of stabilizations that can be applied to each so that the resulting surfaces can be deformed into each other. A consequence is the following:

□ **THEOREM 3.** *Any two Seifert matrices for a knot are S-equivalent.*

□ **COROLLARY 4.** *If V_1 and V_2 are Seifert matrices associated to the same knot, then the polynomials $\det(V_1 - tV_1^t)$ and $\det(V_2 - tV_2^t)$ differ by a multiple of $\pm t^k$.*

PROOF
This is proved by checking the effect of the two basic operations of S-equivalence on the determinant. The first,

multiplying by M and M^t has no effect on the determinant, since $\det(M) = 1$. The second has the effect of multiplying the determinant by t. □

EXAMPLE

In Section 1 the Seifert matrix of the knot illustrated in Figure 6.1 was presented. The Alexander Polynomial of that knot is given by the determinant of the matrix

$$\begin{pmatrix} 2-2t & 1 & 0 & 0 \\ -t & -5+5t & 1-t & 0 \\ 0 & 1-t & 2-2t & -1+2t \\ 0 & 0 & -2+t & -2+2t \end{pmatrix}$$

The determinant of this matrix is $64t^4 - 272t^3 + 417t^2 - 272t + 64$.

EXERCISES

2.1. Compute the Alexander polynomial of the trefoil knot using the Seifert matrices found in Exercise 1 of the previous section.

2.2. Find the Alexander polynomial of the knot discussed in Exercise 1.3, using the Seifert matrix found there.

2.3. Check the calculation of the determinant that gives the Alexander polynomial of the knot in Figure 6.1.

2.4. Compute the Alexander polynomial of the (p,q,r)-pretzel knot, (p, q, and r odd) by using the Seifert matrix found in Exercise 1.4.

2.5. Use the result of Exercise 1.7 to show that the Alexander polynomial of the connected sum of knots is the product of their individual Alexander polynomials.

2.6. The Alexander polynomial of a knot can be normalized so that only positive powers of t appear and the con-

stant term is nonzero. Show that the degree of the resulting polynomial is even. Hint: use the symmetry condition,
along with the fact that $A_K(1)$ is odd. (If $A_K(1)$ is even,
so is $A_K(-1)$ and the knot would have a mod 2 labeling.
Now see Exercise 3.5, Chapter 3.)

2.7. Show that if the determinant of a $2g \times 2g$ Seifert matrix is nonzero, then the Alexander polynomial is degree
$2g$ and has nonzero constant term.

3 The Signature of a Knot, and Other S-equivalence Invariants   In the last section it was seen that any two Seifert matrices for a knot are S-equivalent; that is, a pair of fairly simple operations will transform one

to the other. Because of this many knot invariants can be
defined using the Seifert matrix. This section discusses a
few of them.

DETERMINANT

The determinant of the Seifert matrix can change under
stabilization, and is not an invariant of the knot. However,
if V is the Seifert matrix of a knot, then the determinant of
$V + V^t$ is only changed by a sign if the matrix is stabilized.
This is an easy exercise in determinants, and is given in the
problems below. Multiplying by a matrix of determinant
± 1 can at most change the sign of the determinant as well.
Hence, the absolute value of the determinant of $V + V^t$ is
a well-defined knot invariant.

This is in fact the same as the determinant invariant
defined in Chapter 3. The determinant of $V + V^t$ is the
value of the Alexander polynomial evaluated at $t = -1$ up
to a sign. The Seifert matrix approach leads to a simple
calculation of the determinant.

THE SIGNATURE OF A KNOT

Given a symmetric $(A = A^t)$ real matrix, there is a signature defined. One definition is constructive. By performing a sequence of simultaneous row and column operations the matrix can be diagonalized. The *signature* of the matrix is defined to be the number of positive entries minus the number of negative entries on the diagonal.

EXAMPLE

Consider the symmetric matrix A_1 below. Multiply the first row by $-1/4$ and add it to the second row. Now perform the same operation using the first column. The resulting matrix is listed as A_2.

$$A_1 = \begin{pmatrix} 4 & 1 & 0 & 0 \\ 1 & 10 & 2 & 0 \\ 0 & 2 & 4 & -3 \\ 0 & 0 & -3 & -4 \end{pmatrix}$$

$$\rightarrow \begin{pmatrix} 4 & 0 & 0 & 0 \\ 0 & 39/4 & 2 & 0 \\ 0 & 2 & 4 & -3 \\ 0 & 0 & -3 & -4 \end{pmatrix} = A_2$$

Using the second row and column the nondiagonal entries of the second row and column can be changed to 0. Finally, working with the third column and row reduces the matrix to diagonal form. The exercises ask you to check that the final result is

$$\begin{pmatrix} 4 & 0 & 0 & 0 \\ 0 & 39/4 & 0 & 0 \\ 0 & 0 & 140/39 & 0 \\ 0 & 0 & 0 & -911/140 \end{pmatrix}$$

As there are 3 positive entries and 1 negative entry the signature is $3 - 1 = 2$.

A theorem of algebra, named for James J. Sylvester, states that if the symmetric matrix B is given by $B = MAM^t$, where M is invertible, then the signatures of A and B are equal.

For a Seifert matrix V of a knot K, the matrix $V + V^t$ is symmetric and its signature is called the *signature of K*, denoted $\sigma(K)$.

□ **THEOREM 5.** *For a knot K, the value of $\sigma(K)$ does not depend on the choice of Seifert matrix, and is hence a well-defined knot invariant.*

PROOF
First, note that if Seifert matrices V and W are related by $W = MVM^t$, then $(W + W^t) = M(V + V^t)M^t$. Hence Sylvester's theorem implies that the signature of $(W + W^t)$ is the same as that of $(V + V^t)$. All that is left to check is that stabilization of V does not change the signature of $(V + V^t)$. Proving this is left to the exercises.   □

EXAMPLE
A Seifert matrix V for the knot in Figure 6.1 was given in Section 1. For that V, $V + V^t$ is the matrix discussed in the previous example, and hence the signature of that knot is 2.

Using the Seifert matrix for the trefoil computed in Exercise 1.1 $V + V^t$ is given by

$$\begin{pmatrix} -2 & -1 \\ -1 & -2 \end{pmatrix}.$$

It has signature -2.

The same calculation for the left-handed trefoil gives a signature of 2. Hence, the right and left trefoils are inequivalent knots.

THE SIGNATURE FUNCTION

The signature of a knot can be generalized by using complex numbers. First recall that a complex matrix is called Hermitian if it equals its conjugate transpose. Any Hermitian matrix can be diagonalized performing a sequence of row and column operations. The only change from the diagonalization of real matrices is that if a row is multiplied by a complex number, then, when the corresponding column operation is performed, the column is multiplied by the conjugate of that number. Once diagonalized, the matrix has real entries, (as it equals its conjugate transpose) and the signature of the matrix is given by the number of positive entries minus the number of negative entries. Again, a theorem of linear algebra states that if a Hermitian matrix A is replaced by MAM^* where M is an invertible complex matrix and M^* is its conjugate transpose, the signature is unchanged.

Let V be the Seifert matrix for a knot K and let ω be a complex number of modulus 1. Consider the Hermitian matrix $(1-\omega)V + (1-\omega^{-1})V^t$. The signature of this matrix is called the ω-*signature* of K. Checking that S-equivalent Seifert matrices have the same ω-signature is straightforward; only stabilization remains to be checked. If one thinks of modulus 1 complex numbers as lying on the unit circle in the complex plane, this signature defines a function on the unit circle called the *signature function* of the knot.

Even for 2×2 Seifert matrices, the signature function can be difficult to compute. (See Exercise 3.8.) However, it can sometimes be used to distinguish knots where other methods fail. It also has many theoretical applications.

EXERCISES

3.1. Complete the diagonalization and signature calculations presented in the section.

3.2. Compute the signature of the $(3,5,-7)$-pretzel knot .

3.3. Compute the determinant of the (p,q,r)-pretzel knot.

3.4. For a Seifert matrix V, $\det(V + V^t) \neq 0$. (Why?) Conclude that the signature of a knot is always even.

3.5. Prove that stabilization does not change the signature of a matrix.

3.6. Use Exercise 1.7 to show that the signature of a connected sum of knots is the sum of their signatures.

3.7. Prove that the matrix $(1-\omega)V + (1-\omega^{-1})V^t$ has nonzero determinant for ω of modulus 1 unless ω is a root of the Alexander polynomial. Conclude that the signature function is constant on the circle, except for a finite number of jump discontinuities.

3.8. Compute the signature function for the trefoil and the figure-8 knot.

3.9. Compute the signature of the $(2,n)$-torus knot using Exercise 2.5.

4 Knot Groups and the Alexander Polynomial   In Chapter 5 it was shown how to construct a presentation of a group, starting with a knot diagram. The presentation consists of a set of n variables, and $n-1$ words in the variables (and their inverses.) In this section an al-

gorithm will be presented that computes the Alexander
polynomial of the knot starting with a group presentation
of the form arising from the construction given in Chapter
4. The algorithm was discovered by Fox. It is, in fact,
possible to compute the polynomial using any presentation of the group, but to do this the algorithm has to be
generalized.

That the knot polynomial is determined by the group
of the knot has certain theoretical implications. For instance, as mentioned in Section 2, the link between the
combinatorial and geometric definition of the Alexander
polynomial is provided by this algebra. On the practical
side, Fox's algorithm provides one more means of computing the Alexander polynomial.

FOX DERIVATIVES

There is a procedure for defining the formal partial derivatives of monomials in noncommuting variables. In the
present case these monomials will be the defining words
of the group of a knot. The definition of the derivative
begins with two basic rules, which in turn determine the
derivative in general. Fox proved that these rules yield a
well-defined operation on the set of words. Note that the
derivative of a word will no longer be a single word, but
rather the formal sum of words.

1. $(\partial/\partial x_i)(x_i) = 1$, $(\partial/\partial x_i)(x_j) = 0$, $(\partial/\partial x)(1) = 0$.
2. $(\partial/\partial x_i)(w \cdot z) = (\partial/\partial x_i)(w) + w \cdot (\partial/\partial x_i)(z)$, where
   w and z are words in variables $\{x_j,\ x_j^{-1}\}$.

One immediate consequence is that

$$\frac{\partial}{\partial x_i}(x_i^{-1}) = -x_i^{-1}.$$

This follows from the calculations $(\partial/\partial x_i)(x_i \cdot x_i^{-1}) = (\partial/\partial x_i)(1) = 0$, and, using rule 2, $(\partial/\partial x_i)(x_i \cdot x_i^{-1}) = 1 + x_i(\partial/\partial x_i)(x_i^{-1})$.

EXAMPLE

The partial derivatives of the equation $xyxy^{-1}x^{-1}y^{-1}$ are computed in the following manner. Write the word as $(x) \cdot (yxy^{-1}x^{-1}y^{-1})$ and apply rule 2. To differentiate the second term, write it as $(y)(xy^{-1}x^{-1}y^{-1})$ and use rule 2 again. Proceed in this way, factoring out one term at a time. The final result is that the derivative with respect to x is $1 + xy - xyxy^{-1}x^{-1}$. The derivative with respect to y is $x - xyxy^{-1} - xyxy^{-1}x^{-1}y^{-1}$. In the exercises you are called on to fill in the details of this calculation, and to compute some more complicated examples.

As a hint of things to come, note the following about this example. The equation $xyxy^{-1}x^{-1}y^{-1}$ is the defining equation for the group of the trefoil knot. If in the derivative, $1 + xy - xyxy^{-1}x^{-1}$, the variables are both replaced with t, then the polynomial $1 - t + t^2$ results. This is the Alexander polynomial of the trefoil. (Also, if the substitution is made in $x - xyxy^{-1} - xyxy^{-1}x^{-1}y^{-1}$, the polynomial $-t^2 + t - 1$ results, which is the same as the first modulo a multiple of $\pm t^i$.)

USING THE FOX CALCULUS TO COMPUTE
THE ALEXANDER POLYNOMIAL

Here is a new algorithm for computing the Alexander polynomial of a knot. Take any presentation of the group of the knot found by the procedure outlined in Chapter 5. The presentation will have one more generator than relation. Now form the Jacobian matrix consisting of all the partial derivatives of the equations, and eliminate any one column of the matrix. Substitute t for all the variables that appear. Finally, take the determinant of the matrix that results. This determinant is the Alexander polynomial.

In Chapter 5 it was shown that the group of the knot illustrated in Figure 5.8 was generated by x, y, and z, subject to the relations:

$$r_1 = yx^{-1}zxy^{-1}xyx^{-1}z^{-1}xy^{-1}xy^{-1}x^{-1} = 1,$$
$$r_2 = z^{-1}y^{-1}zyxy^{-1}z^{-1}yzyx^{-1}y^{-1}z^{-1}yxy^{-1} = 1.$$

(Recall that any one of the 3 relations is a consequence of the other 2.) If, in the Jacobian, the column corresponding to $\partial/\partial y$ is eliminated, the resulting matrix is 2×2. As an example, the (1,2) entry is $\partial/\partial z(r_1) = yx^{-1} - yx^{-1}zxy^{-1}xyx^{-1}z^{-1}$. Substituting t for each variable yields $-1 + t$. If the other derivatives are computed and t substituted, the resulting matrix is

$$A(t) = \begin{pmatrix} -t^2 + 4 - 2 & -t + 1 \\ -t + 2 & 1 - 3t^{-1} + t^{-2} \end{pmatrix}$$

Taking the determinant yields an Alexander polynomial $-2t^2 + 10t - 15 + 10t^{-1} - 2t^{-2}$.

WHY THIS WORKS

The proof that this procedure actually produces the Alexander polynomial is fairly long and technical. The basic ideas are easily explained.

To begin, there is the following central observation. One presentation of the knot group is obtained with no algebraic manipulations. For each arc there is a generator and for each crossing there is a relationship. For instance, at a right-hand crossing there is the relation $x_i x_j x_i^{-1} x_k^{-1} = 1$. If the Jacobian matrix for this set of relationships is computed and then t is substituted for all the variables, the resulting matrix is just the matrix used in the combinatorial definition of the polynomial given in Chapter 3. The algebraic manipulations that reduce the number of variables in the presentation correspond to operations on the Jacobian matrix. A careful calculation shows that none of these changes affect the final determinant.

EXERCISES

4.1. The knot 5_1 has knot group

$$\langle x,\ y \mid xyxyxy^{-1}x^{-1}y^{-1}x^{-1}y^{-1} \rangle.$$

Compute its Alexander polynomial.

4.2. Find two generator presentations of the groups of the knots 6_2, 6_3, 7_1, and 7_5. In each case use the presentation to compute the Alexander polynomial.

4.3. Fill in the details of the calculation of the matrix $A(t)$ in this section.

4.4. If a knot diagram has n crossings, there is an n generator presentation of the knot group. Show that if this presentation is used to compute the Alexander polynomial, the result is the same as in the combinatorial calculation in Chapter 3.

CHAPTER 7:
NUMERICAL INVARIANTS

A few methods for associating integers to knots have already appeared in the text. The genus is an important example. Others include the signature, the determinant, and the mod p rank. In this chapter many more will be described. Some of these will seem to be very natural quantities to study. Others, such as the degree of the Alexander polynomial, may at first seem artificial; it is the relationship between these invariants and the more natural ones that is particularly interesting and useful.

It will be clear in this chapter that with the introduction of each new invariant a host of questions arise concerning its relationship with other invariants. Some of these questions will be discussed, others will be presented in the exercises. A few open questions will appear along the way.

1 Summary of Numerical Invariants

Several knot invariants have been defined so far. These are reviewed in this section. In the next sections many new invariants will be described.

GENUS

Every knot forms the boundary of an oriented surface called a Seifert surface of the knot. The genus of a knot, $g(K)$, is the minimal genus that occurs among all Seifert surfaces. Only the unknot has genus 0; the pretzel knots form an infinite family of genus 1 knots. The proof of the prime decomposition theorem was based on the result that genus is additive under connected sum.

Another similar notion of genus is based on nonorientable surfaces. This concept plays a secondary role to orientable genus, and will not be pursued.

MOD p RANK

Finding mod p labelings of a knot diagram can be reduced to solving a system of linear equations mod p. The dimension of that solution space is called the mod p rank of the knot. In Exercise 4.6 of Chapter 3 it was shown that, if K has mod p rank n, then the number of mod p labelings is $p(p^n - 1)$. It follows that mod p rank is additive under connected sum. (See Exercise 1.1.)

DETERMINANT, DET(K)

The determinant was first defined combinatorially. However, the simplest definition is based on Seifert matrices. If V is a Seifert matrix for a knot K, then the determinant of K, $\det(K)$, is the absolute value of the determinant of $V + V^t$. Thus, the determinant of the connected sum of knots is the product of their determinants (see Chapter 6).

SIGNATURE, $\sigma(K)$

The Seifert matrix also provides a means of defining the signature of a knot. If V is a Seifert matrix for K, then $\sigma(K)$ is the signature of $V + V^t$. Signature is additive under connected sum. See Exercise 3.6, Chapter 6, for a proof. (ω-signatures can also be defined using V.)

As shown earlier, the right- and left-handed trefoils have signature -2 and 2, respectively. Hence, the connected sum of the two trefoils, called the *square knot*, has signature 0. Connected sums of square knots provide an infinite family of knots with signature 0.

DEGREE OF THE ALEXANDER POLYNOMIAL

Although not yet discussed, this invariant derives easily from the polynomial itself. By multiplying by the appropriate power of t, the Alexander polynomial of a knot can be normalized to have no negative powers of t, and so that the constant term is nonzero. The degree of this polynomial is called the degree of the Alexander polynomial.

The Alexander polynomial of a connected sum of knots is the product of their individual polynomials (see Chapter 6). Hence, the degree of the Alexander polynomial adds under connected sum. An infinite family of knots, all with Alexander polynomial 1 can be constructed from the connected sums of copies of a single nontrivial polynomial 1 knot. Families containing only prime knots also exist.

EXERCISE

1.1. If a knot K has mod p rank n, then the number of mod p labelings is $p(p^n - 1)$. Use this to show that the number of labelings including ones with all labels the same is given by p^{n+1}. Use this to prove that mod p rank adds under connected sum.

---

2 New Invariants   The two invariants defined in this section are the most natural in the study of knots. Surprisingly, although they are so simple to define their calculation turns out to be especially difficult, and the most natural questions concerning them are unanswered.

CROSSING INDEX, $C(K)$

Each regular projection of a knot has a finite number of double points. Different projections of a knot can have different numbers of double points, since Reidemeister moves 1 and 2 change the number of double points. The least possible number of double points in a projection of a knot is called the crossing index of the knot.

For example, the unknot has crossing index 0. It is fairly easy to see that if a knot has a projection with one or two crossings it is unknotted. Hence there are no knots of crossing index 1 or 2. The trefoil has crossing index 3.

Although there are clearly only a finite number of knots with a given crossing index, listing them all is difficult. The chart of prime knots in the appendix is arranged by crossing index. The number of knots of a given crossing index seems to grow very rapidly, but little is known in detail about this number.

At the present time it is conjectured, but unproven, that the crossing index adds under connected sum. (This has been proved for knots with alternating projections; a knot diagram is alternating if, travelling around the knot, overpasses and underpasses are met alternately. This result for alternating knots is discussed again in Chapter 10.) As a measure of the present state of ignorance, we cannot rule out the possibility that the connected sum of two knots can have crossing number less than either factor!

UNKNOTTING NUMBER, $U(K)$

Given a knot diagram, it is always possible to find a set of crossings such that if each is switched from right- to left-handed or vice versa, the knot becomes unknotted. One way to discover one set of such switches is to draw a new knot diagram starting with the projection of the knot. Trace the knot *projection* starting at a point p. Each crossing point will be met twice in the tracing, and when it is met for the second time, have that strand go under the first. This is best understood via an example; the result of this construction for a particular knot is illustrated in Figure 7.1. The proof that the algorithm produces an unknot is left to the exercises.

Figure 7.1

For a given knot diagram several different choices of crossing change can lead to the unknot, and the number of crossing changes that are required might depend on the choice of diagram. The minimal number of crossing changes that is required, ranging over all possible diagrams, is called the unknotting number of the knot.

Given that the definition is taken over all possible diagrams, the unknotting number seems difficult to compute, and in general it is. However, only the unknot has unknotting number 0. The n-twisted doubled knots considered in Exercise 2.2 of Chapter 3 (see also Exercise 1.3 of Chapter

6) provide an infinite family of unknotting number 1 knots.
They are distinguished by their Alexander polynomials.

How the unknotting number behaves under connected
sums is a mystery. It is easily proved that the unknotting
number of the connected sum of knots is at most the sum
of their unknotting numbers, and the conjecture is that
unknotting number is additive. Scharlemann has proved
that the connected sum of two unknotting number one
knots is always of unknotting number two.

A fascinating example concerning the unknotting
number was discovered by S. Bleiler. Figure 7.2 presents
two diagrams of the same knot, the second with more crossing than the first. No two crossing changes in the first
diagram produces an unknot, but changing the indicated
crossings in the second diagram does unknot it.

Figure 7.2

Bleiler proved that to demonstrate that the knot has unknotting number 2 the crossing number of the diagram
used cannot have the minimal number of crossings for the
knot. (Figure 7.2 presents only one minimal crossing diagram of the knot; there conceivably could be more.) The
next section includes the needed techniques to prove that
this knot has unknotting number ≥ 2.

EXERCISES

2.1. Draw all knot diagrams having 2 crossings.

2.2. Prove that there are only a finite number of n-crossing knots for each integer n.

2.3. Prove that the procedure outlined in the text actually produces an unknotted curve.

2.4. Check that making the indicated crossing changes in Bleiler's example (Figure 7.2) produces the unknot. Show that no two crossing changes in the first diagram gives the unknot.

3 Braids and Bridges Although somewhat less intuitive then the crossing index and the unknotting number, both of the invariants described in this section have a long history in the study of knots. The study of braids is particularly fascinating in that it introduces group theory into the study of knots in a completely new way.

Figure 7.3

BRAIDS

An n-stranded braid consists of n disjoint arcs running vertically in 3 space. The set of starting points for the arcs must lie immediately above the set of endpoints. Figure 7.3 illustrates a 5-braid. A formal definition need not be given, and could be supplied by the reader.

A braid can be turned into a link by attaching arcs to the

top and bottom, as illustrated in Figure 7.4. Braids are of interest in the study of knots and links because of a theorem that states that every knot and link arises from a braid in this way. The proof is constructive, as follows.

Draw the knot polygonally, and orient it. Also pick a point in the projection plane which does not lie on the knot. This point will be called the *braid axis*. The goal of the construction is to arrange for every segment of the polygon to run clockwise with respect to the chosen point. If some segment runs counter clockwise, it can be divided up into several smaller segments, each of which can be pulled across the axis. This is illustrated in Figure 7.5. Exercise

Figure 7.4

2 asks that you apply this algorithm to several knots to draw them as closed braids.

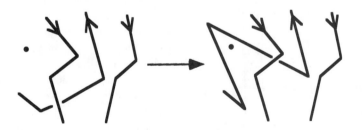

Figure 7.5

Different braids can close to form the same knot; the *braid index* of a knot, denoted brd(K), is defined to be the minimum number of strands that are required in a braid description of a knot. Braid index is subadditive under connected sum; that is, brd($K\#J$) $\leq$ brd(K) + brd(J). To see this, note that given braid descriptions of two knots, there is a simple way to construct a braid description of their connected sum. This is illustrated in Figure 7.6.

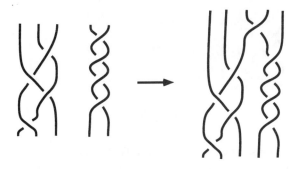

Figure 7.6

Artin introduced braids into the study of knots. What is most fascinating about braids is that there is a natural way to form groups using them. Given two n-stranded braids, placing one on top of the other produces a new braid. This operation induces a group operation on the set of equivalence classes of n-stranded braids, where two braids are equivalent if one can be deformed into the other fixing all endpoints. In the exercises you are asked to derive a few properties of this group, called the *braid group*.

One important theorem in the study of braids deserves notice. As was mentioned, two distinct braids can produce the same knot or link when closed up. For instance, stabilization, as indicated in Figure 7.7, does not effect the resulting link. Also, if a given braid is multiplied on the right and left by a second braid and its inverse (in the braid group) the resulting links are the same. This operation is called conjugation in the braid group. A theorem

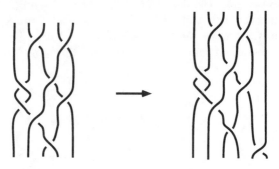

Figure 7.7

of Markov states that if two braids give the same knot or link, then each can be repeatedly stabilized and conjugated so that the same braid results. This theorem, along with a knowledge of the structure of the braid group, was crucial for Jones' discovery of new polynomial invariants of knots. More on that later.

BRIDGE INDEX, $\mathrm{brg}(K)$

Any projection of a knot can be perturbed so that there are a finite number of relative maxima. Figure 7.8 illustrates a knot with the maxima and minima marked. You can prove that the number of minima equals the number of maxima. Different diagrams of a knot can certainly have a different number of maxima. The minimum number of such maxima (taken over all possible projections) is called the *bridge index* of the knot, denoted $\mathrm{brg}(K)$.

It should be clear that only the unknot has bridge index 1. Hence the bridge index of the trefoil is two, as can be seen in its standard projection.

The first 3-bridge knot in the table of prime knots is 8_5. A theorem proved by Schubert states that the bridge index behaves nicely under the connected sum operation.

Figure 7.8

□ **THEOREM 1.** *For knots K and J, $\operatorname{brg}(K\#J) = \operatorname{brg}(K) + \operatorname{brg}(J) - 1$.*

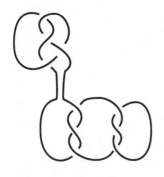

Figure 7.9

The proof is quite difficult. One step is demonstrated easily in a diagram; the bridge index satisfies the inequality $\operatorname{brg}(K\#J) \leq \operatorname{brg}(K) + \operatorname{brg}(J) - 1$. Figure 7.9 illustrates the connected sum of a 2-bridge knot and a 3-bridge knot drawn so that it has 4 bridges.

A simple corollary of the Schubert theorem is that 2-bridge knots are prime (See Exercise 3.3.) Even this is a difficult geometric exercise without the aid of Schubert's general result.

EXERCISES

3.1. The n stranded braid group is generated by the twists σ_i which put a half twist between the i-th and $(i+1)$-th strand, as indicated in Figure 7.10 below. Show that the two relations hold: $\sigma_i\sigma_{i+1}\sigma_i = \sigma_{i+1}\sigma_i\sigma_{i+1}$, and $\sigma_j\sigma_i = \sigma_i\sigma_j$, $|i-j| > 1$. (In fact, these two sets of relations generate all the relations in the braid group.)

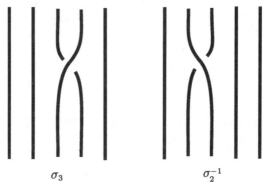

σ_3 σ_2^{-1}

Figure 7.10

3.2. Draw the knots 4_1 and 5_2 as closed braids.

3.3. How does Theorem 1 imply that 2-bridge knots are prime?

3.4. Any 2-bridge knot can be drawn with one strand straightened and not crossing any of the other strands, as illustrated in Figure 7.11 below. Describe a method for converting a 2-bridge diagram into this form. (With this observation the classification of 2-bridge knots can be stated. Any 2-bridge knot is determined by a sequence of integers, $[c_1, c_2, ..., c_n]$, where c_i is the number of right- or left-handed twists, depending on i odd or even.)

The knot illustrated to the left corresponds to $[2,2,3]$. To such a sequence one can form the continued fraction,

$$\frac{p}{q} = c_1 + \cfrac{1}{c_2 + \cfrac{1}{c_3 + \cdots}}.$$

Now Schubert proved that two 2-bridge knots, with corresponding fractions p/q and p'/q', are equivalent if and only if $p = p'$ and $q - q'$ is divisible by p.)

Figure 7.11

3.5. Apply your algorithm from Exercise 3.4 above to illustrate the knots 7_3 and 8_2 in standard form. What are the associated fractions for each?

3.6. How does the continued fraction corresponding to a 2-bridge knot compare to that of its mirror image? Which two bridge knots are equivalent to their mirror images?

4 Relations between Numerical Invariants

Many of the numerical invariants studied so far are closely related. For instance, the combinatorial algorithm for computing Alexander polynomials immediately implies that the degree of the Alexander polynomial is less than the crossing number. Hence, the $(2, n)$-torus knot cannot be drawn with fewer than n crossings; the degree of its polynomial was discussed in Chapter 3, Section 5, and shown to be $n - 1$. This section will focus on demonstrating a few of the less obvious connections.

The next section will deal with the independence of some of the invariants.

THE CROSSING NUMBER AND THE GENUS

Recall that Seifert's algorithm provides a means of building a Seifert surface for a knot from its diagram. In Exercise 3.4 of Chapter 4, it was show that the genus of the resulting Seifert surface is given by $2g = cr - s + 1$, where cr is the crossing number of the diagram and s is the number of Seifert circles. Unless K is unknotted, $s > 1$, so $2g \leq cr - 1$. For the trefoil knot, $2g = cr - 1$.

BRIDGE INDEX AND MOD p RANK

Any mod p labeling of an n-bridge knot is determined by the labels on the n top arcs, or bridges. Hence, there can be at most an n-dimensional space of labelings. Taking into account the 1-dimensional space of trivial labelings, one has that the mod p rank of a knot is at most the $\mathrm{brg}(K) - 1$. As an application, the $(3,3,3)$-pretzel knot has mod 3 rank 2, and so cannot be drawn with 2 bridges. It is clearly a 3-bridge knot.

SIGNATURE AND THE UNKNOTTING NUMBER

Arguments concerning the unknotting number are much more difficult. The result here states that $2u(K) \geq |\sigma(K)|$. The proof depends on showing that changing a crossing in a knot changes the signature by at most 2.

Fix a knot diagram and a crossing in the diagram. If Seifert's algorithm is applied to the diagram the resulting Seifert surface is built from many disks and the given crossing corresponds to a band joining two of the disks. To find the Seifert matrix the surface must be deformed into a single disk with bands added. For the calculation this must be done in such a way that the given band corresponds to a single band on the final surface.

To see that this is possible, cut the Seifert surface across the band of interest. The remaining surface can be assumed to be connected. (Why?) Deform it into a single disk with bands added. The original Seifert surface can be recovered by reattaching the band that was cut to the disk. Order the bands so that this final band is the last in the ordering.

Changing the crossing of interest will have the effect of twisting the last band. This will in turn only effect the last diagonal entry of the Seifert matrix, V. Hence, the diagonalization of $V + V^t$ only changes in its last entry, and the signature can change by at most 2. The signature of Bleiler's example is 4, and this is how he proves it does not have unknotting number 1.

MOD p RANK AND UNKNOTTING NUMBER
In general the unknotting number is at least as large as the mod p rank, for all p. All that will be proved here is that unknotting number 1 knots have mod p rank ≤ 1. The reader should interpret the statement and argument in terms of colorings. (Colorings are often used in expository talks on knot theory to prove that the trefoil is not unknotted. The following argument translates into an easy proof of the much subtler fact that the square knot cannot be unknotted with a single crossing change, regardless of how it is drawn.)

Suppose that a knot K has unknotting number 1, and fix a diagram for K and the crossing which changes K into an unknot when reversed. If there is a nontrivial labeling of K for which both the over and undercrossings are labeled 0 a contradiction arises. The given labeling remains consistent when the crossing is changed, yielding a nontrivial labeling of the unknot.

If the knot has mod p rank > 1, then there are two linearly independent labelings, both of which are 0 on the overcrossing. Neither can be 0 on the undercrossing by the previous argument. However subtracting some multiple of one labeling from the other yields a labeling with the bottom label 0. (Recall that the multiple is taken mod p.) The new labeling is nontrivial by linear independence.

EXERCISES

4.1. Prove that for any knot K, the degree of the Alexander polynomial is at most twice the genus.

4.2. Prove that the $|\sigma(K)| \leq 2g(K)$.

4.3. (a) Prove that the bridge index of a knot is at most equal to the braid index.

(b) Find an example of a 2-bridge link that has braid index greater that 2. (Linking numbers should help here.) Find a similar example of a knot.

4.4. (a) Prove that for n even, an n-crossing knot has genus at most $(n-2)/2$.

(b) Prove that if K has crossing number n, with n odd, then either K is a $(2,n)$-torus knot, or K has genus at most $(n-3)/2$. (The torus knot has genus $(n-1)/2$.)

---

5 Independence of Numerical Invariants While some numerical invariants are closely related, others are completely independent. In most cases, this is demonstrated by constructing families of examples. Some of the families of examples are constructed from a few basic examples and connected sums. Others are much more complicated. Here a few will be surveyed, with the main focus on bridge index.

BRIDGE INDEX AND THE DEGREE OF THE ALEXANDER POLYNOMIAL

There is no relationship between the degree of the Alexander polynomial and the bridge index of a knot. The $(2,n)$-torus knots provide examples of two bridge knots with arbitrarily high degree Alexander polynomial. On the other hand, by forming the connected sum of many polynomial 1 knots, a polynomial 1 knot with large bridge index is created.

INDEPENDENCE OF *mod p* RANKS

The trefoil knot has mod 3 rank 1 and mod 5 rank 0; the $(2,5)$-torus knot has mod 3 rank 0 and mod 5 rank 1. Hence, the connected sum of k trefoils and j 5-twist knots has mod 3 rank k and mod 5 rank j. It follows that in general there is no relationship between the mod 3 and mod 5 ranks.

Given any finite set of primes, similar examples can be constructed showing the independence of mod p ranks. Note that it is not possible to find a knot with specified mod p ranks for all primes. For a given knot only a finite number of the mod p ranks are positive. The determinant of a knot provides a bound on the number of primes p for which the mod p rank can be positive. Exercise 5.1 asks for a precise bound.

SIGNATURE AND BRIDGE INDEX

The $(2,n)$-torus knot knot has signature $n-1$, and is a two-bridge knot. (See Exercise 3.9, Chapter 6) Hence no bound on the signature can be based on the bridge index. On the other hand, the connected sum of square knots has 0 signature, but large bridge index, so no bound on the bridge index follows from the signature.

UNKNOTTING NUMBER AND THE BRIDGE INDEX
The $(2,n)$-torus knots give a family of 2-bridge knots with
arbitrarily high unknotting number. (Consider the signature.) The process of doubling a knot, as illustrated in
Figure 7.12, produces unknotting number 1 knots of large
bridge index.

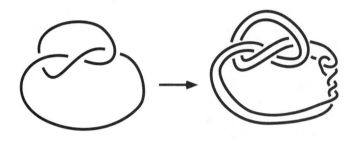

Figure 7.12

Schubert proved that if a knot is doubled the bridge
index of the resulting knot is twice that of the original
knot, except in one special case. (See Exercise 5.3.) It is
clear that the bridge index of a doubled knot is at most
twice that of the original knot, but showing that there is
an equality is a lengthy and delicate geometric argument.

Without that delicate geometry, it is possible to prove
that certain doubled knots have high bridge index, using
the algebraic methods of Chapter 5, specifically labelings
from the symmetric group, S_n. One part of the argument
is based on the following theorem.

☐ **THEOREM 2.** *If a knot K can be labeled with transpositions from S_n then $\mathrm{brg}(K) \geq n$.*

PROOF

Given such a labeling of K, the set of labels generates S_n. However, the labels on the bridges determine all the other labels, as was seen in Chapter 5. Hence, the labels that occur on the bridges must generate S_n. According to Exercise 1.8 of Chapter 5, S_n cannot be generated by fewer than $n-1$ transpositions. The result follows.   □

To apply this to the construction of examples, suppose that one starts with a knot diagram that has been consistently labeled with 3-cycles from S_n. (It is not required, or for that matter even possible, for the labels to generate S_n.) This labeling leads to a consistent labeling of some double of the knot using transpositions, as follows: On the bridges of the knot, if the original arc was labeled with the 3-cycle (a,b,c), label the two strands with (a,b) and (a,c). The consistency condition leads to a labeling of the rest of the doubled knot. Any problem with consistency at the bottom can be cured by adding twists.

It may not be immediately clear why a consistent labeling occurs in general. The following observations should clarify the situation. The two transpositions on a parallel pair of strands on a bridge were chosen so that their product is the 3-cycle with which the original strip was labeled. When the consistency condition is used to determine the rest of the labels, this property for adjacent pairs of labels is true everywhere. That is, the labels on any parallel pair of arcs have product equal to the 3-cycle that the original arc of the knot was labeled with. It is now easily checked that along the bottom strands, if the labels do not match up, twists can be added to the pair of strands so that they do match.

The discussion above shows how, given a knot which is consistently labeled with 3-cycles from S_n, it is possible to produce some double of the knot which can be consistently labeled with transpositions from S_n. These transpositions will generate S_n if the original set 3-cycle labels formed a *transitive* set. (A set of permutations is called transitive if for every positive integer $i \leq n$, some product of elements in the set maps 1 to i.) The proof of this algebraic condition is left to the reader as another exercise concerning the symmetric group. The construction is completed by noting that the connected sums of k (2,5)-torus knots can be consistently labeled with a transitive set of 3-cycles from S_{3+2k}. Hence, an explicit example is constructed by forming the connected sum of k (2,5)-torus knots, consistently labeled with a transitive set of 3-cycles from S_{3+2k}.

GENUS AND THE BRIDGE INDEX

The (2,n)-torus knots provide examples of 2-bridge knots of arbitrarily high genus. On the other hand, doubled knots have genus 1. Figure 6.5 illustrates a genus one surface bounded by a double of the unknot; the right-hand band on that surface can itself be knotted so that the resulting surface forms a genus 1 Seifert surface for an arbitrary doubled knot. It was just shown that doubled knots can have arbitrarily large bridge index.

EXERCISES

5.1. The number of primes for which a knot can have nontrivial mod p labelings is bounded by a function of the determinant. Find one such bound.

5.2. Why do doubled knots all have unknotting number 1?

5.3. Find the example of a double of a knot for which the bridge index is not twice the bridge index of the original knot.

5.4. Check the details of the construction of the labeling of a doubled knot with transpositions, given a 3-cycle labeling of the knot being doubled. In particular, check that consistency can be assured by adding the appropriate twists at the bottom.

5.5. Show that the connected sum of k $(2,5)$-torus knots can be labeled with 3-cycles from S_{3+2k} so that the set of labels form a transitive set.

5.6. Figure 7.13 illustrates a genus 3 Seifert surface. Show that its boundary has unknotting number 1. Show that its Alexander polynomial is of degree 6, and hence the knot is exactly genus 3. Generalize this example to find unknotting number 1 knots of arbitrarily large genus. It is more difficult, but possible, to show that there are genus 1 knots of high unknotting number.

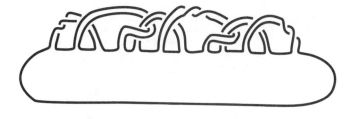

Figure 7.13

CHAPTER 8:
SYMMETRIES OF KNOTS

Knot diagrams can appear symmetrical, and for those that do not, the lack of symmetry is often an artifact of the diagram, and is not inherent in the knot itself. For instance, Figure 8.1 presents two diagrams for the knot 7_6. The first shows no apparent symmetry, while the second is quite symmetrical; a rotation of 180 degrees about a point in the plane leaves the diagram unchanged. As the example indicates, finding symmetrical diagrams for a knot can be a challenging task. On the other hand, powerful tools are available for proving that a knot does not have hidden symmetries.

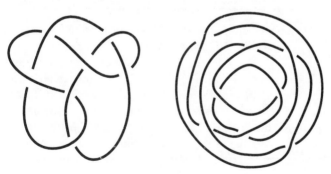

Figure 8.1

Section 1 expands on some of the basic types of symmetry discussed earlier. (For example, it was shown that the trefoil is distinct from its mirror image using the signature; the relationship between a knot and its mirror image will be discussed further.) The rest of the chapter is devoted to another type of symmetry, *periodicity*; roughly stated, a knot is called periodic if it has a diagram that is carried back to itself when rotated about the origin; Figure 8.1 shows that 7_6 is periodic, with period 2.

The two main results of the chapter are theorems of Murasugi and Edmonds. The first places algebraic restrictions on the Alexander polynomials of periodic knots. The second restricts their Seifert surfaces. Together these two theorems provide powerful means for studying the periods of knots. The examples in the final section will demonstrate the beautiful and subtle interplay between geometry and algebra.

1 Amphicheiral and   Given an oriented knot, K,
Reversible Knots   reversing the orientation creates a new oriented knot called its reverse, and denoted K^r. Changing all of its crossings yields an oriented knot denoted K^m. In Chapter 2, Exercise 5.6 asked you to prove that changing the crossings in a diagram for K yields a knot equivalent to the mirror image of K, corresponding to the reflection of its diagram through the y-axis of the knot diagram.

☐ **DEFINITION.** *An oriented knot K is called reversible if K is oriented equivalent to K^r. It is called positive amphicheiral if it is oriented equivalent to K^m, and negative amphicheiral if it is oriented equivalent to K^{rm}.*

EXAMPLES

Figure 8.2 illustrates that a 180 degree rotation about the y-axis carries the knot 4_1 (the figure-8 knot) to itself, but reverses its orientation. Hence it is reversible. The reader should have no trouble showing that if all the crossings are changed, the resulting knot can be deformed to appear again as in the diagram. This shows that the figure-8 is amphicheiral, and, since it is reversible, it is both positive and negative amphicheiral. (See Exercise 1.1.)

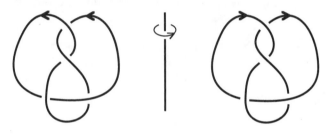

Figure 8.2

Figure 8.3 illustrates the $(3,5,3)$-pretzel knot. It too is reversible; rotate it 180 degrees about the vertical axis in

Figure 8.3

the diagram. It is now known that the only reversible pretzel knots are those with two of the bands having an equal number of twists. A signature calculation shows that this pretzel knot is neither positive nor negative amphicheiral. (It follows from Exercise 1.8 of Chapter 6 that the signature of a knot and its mirror image are negatives.) Finding knots that display one, but not both, forms of amphicheirality is at least as difficult as constructing nonreversible knots.

STRONG SYMMETRY
Although reversible and amphicheiral knots contain symmetries, the symmetry may be hidden. That is, it may be the case that the symmetry cannot be displayed in a diagram. In particular, some knots are reversible, but the reversal cannot be carried out in a simple manner as in the previous examples.

☐ **DEFINITION.** *A knot is called strongly reversible if it is equivalent to a knot that is carried to its reverse by either a 180 degree rotation about the y-axis, or reflection through the (y, z)-plane.*

If the standard diagram for the $(3, 5, 3)$-pretzel knot is rotated by 180 degrees about the y-axis, then the representative for the knot is clearly fixed. On the other hand, the connected sum of the left- and right-handed trefoils (see Figure 8.4) is not invariant under that rotation; it clearly is invariant when reflected through the (y, z)-plane.

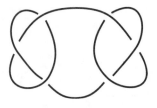

Figure 8.4

It was once conjectured that a reversible knot is necessarily strongly reversible. This is now known to be false. The double of a knot is always reversible, as the reversal can be carried out inside a torus, as illustrated in Figure

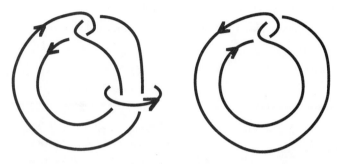

Figure 8.5

8.5. However, Whitten proved that for a double of a knot to be strongly reversible, the original knot itself has to be reversible. The proofs depend on difficult geometric constructions.

There are also similar notions of strong amphicheirality. A knot K is called *strongly positive amphicheiral* if there is a self-map T of 3-space with $T^2 = $ identity, such that $T(K) = K^m$. Similarly K is called *strongly negative amphicheiral* if there is such a T with $T(K) = K^{rm}$. As our only example, the connected sum $K \# K^{rm}$ is strongly negative amphicheiral. Such a connected sum is illustrated in Figure 8.4. Let T be rotation by 180 degrees about the y-axis. The effect of T is the same as changing all the crossing in the diagram. As with reversibility, examples exist demonstrating the distinction between the various notions of amphicheirality.

EXERCISES

1.1. Prove that for reversible knots, being positive amphicheiral is equivalent to being negative amphicheiral.

1.2. (a) Verify that 6_3 is amphicheiral.

(b) Show that 6_3 is reversible.

1.3. Verify that the second knot in Figure 8.1 is 7_6.

2 Periodic Knots   For any integer $q \geq 2$, let R_q denote the linear transformation of R^3 consisting of a rotation about the z-axis of $360/q$ degrees. For any knot K, the diagrams for K and $R_q(K)$ differ by a rotation of $360/q$ degrees about the origin.

☐ **DEFINITION.**   *A knot K is called periodic with period q if K has a diagram which misses the origin and which is carried to itself by a rotation of $360/q$ degrees about the origin.*

Figure 8.6

The diagram in Appendix 1 for the trefoil, 3_1, displays its 3-fold symmetry; the trefoil is periodic of period 3. Similarly, the diagrams of 5_1 and 7_1 show that they have periods 5 and 7, respectively. Figure 8.6 is another diagram of 5_1, showing that it is also a period 2 knot. The first diagram in the chapter, Figure 8.1, displayed 7_6 as a period 2 knot, although no symmetry at all is evident in the figure in Appendix 1. The reader should scan through the appendix and identify the clearly periodic diagrams.

THE QUOTIENT KNOT AND LINKING NUMBERS

Given a periodic diagram for a knot, there is a simple procedure for constructing a simpler knot, called the *quotient knot*.

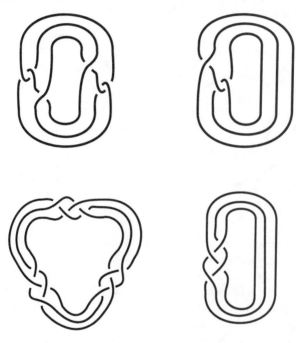

Figure 8.7

In Figure 8.7, two periodic knots and their quotients are drawn. Knots of period 2 and 3 are drawn on the left. Their respective quotients are drawn on the right. Note that for the first the quotient is itself unknotted, and for the second the quotient is the Figure-8 knot.

The construction just given can be reversed: given a knot diagram that misses the origin and an integer $q \geq 2$, one can construct a knot, or link, having the original knot as a quotient. Figure 8.8 illustrates a case in which this so-called *covering link* has more than one component. Deciding whether or not the covering link is a knot calls for the introduction of linking numbers into the study.

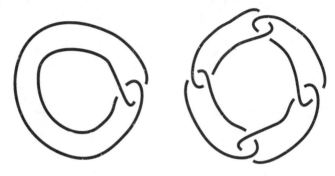

Figure 8.8

Given a diagram for a knot which misses the origin, choose an orientation. Also, pick a ray from the origin such that none of the points of intersection of the ray and knot are tangential. (For a polygonal knot, choose the ray so that it misses all the vertices of the knot.) The *linking number* of the diagram with the z-axis, to be denoted λ, is computed as the absolute value of the intersection number of the knot with the ray. The intersection number is the number of intersection points at which the knot crosses the ray in the clockwise direction minus the number of counterclockwise intersections. For the knot diagram in Figure 8.1, $\lambda = 5$. For the knots in Figure 8.7, the linking numbers are $\lambda = 1$ and $\lambda = 3$. For the knot in Figure 8.8, $\lambda = 0$.

If a knot diagram is periodic, it is easily seen that the linking number of the knot with the z-axis is the same as the linking number of the quotient with the z-axis. (See Exercise 2.5.) Conversely, if a periodic diagram for a knot arises from the covering construction, the linking numbers are the same. It remains to determine when the covering link is a knot.

☐ **THEOREM 1.** *If a knot diagram for K misses the origin, the corresponding q-fold covering link L has a single component if the linking number is relatively prime to q. More generally, the number of components in L is the greatest common divisor of the linking number λ and q.*

PROOF
Observe that neither changes in crossings nor deformations that do not cross the origin affect the linking number or the number of components in the cover. Such deformations determine periodic deformations of the covering link (these are called *lifts* of the deformation on the quotient), and crossing changes clearly have no effect on the algorithm that computes the linking number.

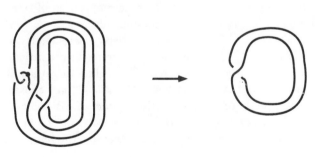

Figure 8.9

Now, by an appropriate
sequence of crossing changes
and deformations, the knot
diagram can be transformed
into one that runs monotoni-
cally around the axis. Crossing changes are used to eliminate any clasps that occur.
This is illustrated in Figure
8.9; after changing the indicated crossing, a deformation
(that does not cross the origin) results in a knot diagram
that runs clockwise about the
origin. Denote the new knot by K'.

Figure 8.10

Pick a ray from the origin meeting the knot in λ
points, and label the points with integers from 1 to λ.
Given any point of intersection on the ray, a new point is
determined by travelling once around the origin along K'.
Hence, a permutation ρ in S_λ is determined. For the knot
illustrated in Figure 8.10, $\rho = (13452)$.

Next observe that as K' is connected, the corresponding permutation is a λ-cycle. In general, K' would have
1 component for each cycle in a decomposition of ρ as a
product of disjoint cycles, including 1-cycles.

The cover of K', say L', similarly corresponds to a
permutation, ρ', and it is easily seen from the construction
that $\rho' = \rho^q$. Now if q is relatively prime to λ then the
q-th power of a λ-cycle is again a λ-cycle. More generally,
the q-th power of a λ-cycle is the product of d disjoint λ/d
cycles, where d is the greatest common divisor of q and λ.
Proving this is one more exercise concerning the symmetric
group.   □

Note that different periodic diagrams of a given knot can have different linking numbers. The trefoil has a periodic diagram of period 3 and linking number 2. It also has a periodic diagram of period 2 and linking number 3, as is shown in Figure 8.11. (A consequence of results of the next section imply that, for a given knot, any two diagrams of the same period also have the same linking number.)

Figure 8.11

EXERCISES

2.1. Figure 8.1 shows that 7_6 can be described as the closure of the square of a 5-strand braid. Show that the same is true for 6_3. The resulting periodic diagram of 6_3 will have 8 crossings.

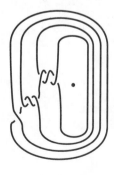

Figure 8.12

2.2. Find 2 crossing changes that convert the knot illustrated in Figure 8.12 into a braid about the origin.

2.3. The braid that results from the crossing changes in Exercise 2 determines a cyclic permutation. Find it.

2.4. Does the statement of Theorem 1 hold when the linking number is 0? Recall that the greatest common divisor of 0 and q is q.

2.5. In the definition of period, it was required that the knot diagram misses the origin. Why is this relevant only in the case of period 2?

2.6. Show that the linking number of a periodic knot with the z-axis is the same as the linking number for the quotient knot.

3 The Murasugi Conditions   Murasugi gave simple but powerful criteria for testing a knot for possible periods; these criteria were based on the Alexander polynomial. He discovered that if a knot has a periodic diagram, then the Alexander polynomial of the knot and its quotient are closely related.

Suppose that a knot K has period $q = p^r$, with p prime. Let J denote the quotient knot of a period q diagram of K, and let λ be the linking number of J with the axis.

☐ **THEOREM 2.** (*Murasugi Conditions*) (1) *The Alexander polynomial of J, $A_J(t)$, divides the Alexander polynomial of K, $A_K(t)$.*

(2) *The following* mod p *congruence holds for some integer i:*

$$A_K(t) = \pm t^i (A_J(t))^q (1 + t + t^2 \cdots + t^{\lambda-1})^{q-1} \pmod{p}.$$

PROOF

The proof of these congruences consists of a lengthy and clever argument in matrix manipulation. Although the details cannot be presented, the idea is fairly simple.

To compute the Alexander polynomial one begins with a labeling of the knot diagram. If the diagram is periodic the labeling can also be chosen to be periodic. For example, if in the quotient knot an arc is labeled x_i, in the covering knot the various lifts of that arc can be labeled $x_i^1, x_i^2, \ldots, x_i^q$. Hence, the corresponding Alexander matrix decomposes into blocks corresponding to the sets $\{x_i^1\}, \{x_i^2\}, \ldots, \{x_i^q\}$. The individual blocks are closely related to the Alexander matrix of the quotient knot. It is perhaps not surprising that the determinant of the large matrix is related to the q-th power of the determinant of the quotient knot. The details of the proof consist of a careful study of the relationship.   □

Figure 8.13

One comment about the second condition offers a little insight. The simplest construction of a period q, linking number λ, knot with quotient J is given by lifting the diagram in Figure 8.13. The covering knot consists of a (q, λ)-torus knot with q copies of J added on. Condition 2 states that any period q knot with the same quotient and linking number has the same polynomial as this basic example, modulo p. Essentially, changes in the diagram of the quotient only change the polynomial of the covering by multiples of p.

PERIODIC KNOTS AND MURASUGI'S CONDITIONS

To begin, a few examples of periodic knots are presented
to demonstrate Theorem 2. The trefoil has a period 3
diagram with quotient the unknot and $\lambda = 2$. Condition 1 is automatically satisfied. Condition 2 implies that
the polynomial of the trefoil, $t^2 - t + 1$, should be equal to
$(t+1)^2 \pmod 3$. Since $2 = -1 \pmod 3$ this congruence
holds.

The period 2 diagram of 7_6 in Figure 8.1 has quotient
the unknot and $\lambda = 5$. Again, Condition 1 is immediate.
Condition 2 states that the Alexander polynomial of 7_6
should be congruent to $t^4 + t^3 + t^2 + t + 1 \pmod 2$. The
polynomial is in fact $t^4 - 5t^3 + 7t^2 - 5t + 1$, so the congruence holds.

EXAMPLES OF CONSTRAINTS ON THE PERIOD OF A KNOT

Theorem 2 provides a powerful means of proving that a
knot does not have certain periods. Before presenting examples, two comments concerning polynomials with mod p
coefficients are needed. First, if p is prime then polynomials factor uniquely mod p into irreducible polynomials. In
this setting uniqueness means that the factors in any two
factorizations can be paired so that each pair differs by at
most multiplication by a constant. Second, if polynomials
f and g have mod p degrees d_1 and d_2 respectively, then
their product has mod p degree $d_1 + d_2$. Here the mod p
degree is the highest degree term with coefficient not divisible by p. Both these facts are proved in introductory
texts in algebra.

The following arguments depend on apparently *ad hoc*
degree calculations. The exercises will develop and clarify
the procedures.

As a first application, the only periods of the trefoil
are 2 and 3. First, suppose that it has period $q = p^r$, with

p prime. If $q > 3$ then Condition 2 quickly implies that either $A_K(t)$ has degree greater than 2, or degree 0, neither of which is the case. To deal with composite powers, note that if a diagram for a knot is of period q, it is also of period q' for all divisors q' of q. The only case that remains for the trefoil is period 6. But a period 6 diagram for the trefoil is also a period 3 diagram, and using Condition 2 one can conclude that $\lambda = 3$. At the same time, it would be a period 2 diagram for the trefoil, and Condition 2 would imply that $\lambda = 2$, contradicting the previous calculation.

As another example, consider the knot 9_{42}, with polynomial $A_K(t) = t^4 - 2t^3 + t^2 - 2t + 1$. The following argument shows that Theorem 2 implies that 9_{42} has no periods. Note that to show this it is sufficient to prove that 9_{42} has no prime periods, p.

Degree considerations arising from Condition 2 imply that $p \leq 5$. For $p = 5$, degree considerations imply $\lambda = 2$ and that $A_J(t)$ has mod 5 degree 0. Condition 2 cannot be satisfied even in this case, as $A_K(t) \neq (t+1)^4 \pmod 5$.

The primes 2 and 3 require individual attention. For $p = 3, A_K(t) = t^4 + t^3 + t^2 + t + 1 \pmod 3$, which is irreducible; it is easily checked that $t^4 + t^3 + t^2 + t + 1$ has no mod 3 roots, and hence no linear factors in its mod 3 factorization, and a more careful check shows that it has no quadratic factors in a mod 3 factorization. Condition 2 then applies to show that the knot cannot have period 3.

To check period 2, note that

$$A_K(t) = (t^2 + t + 1)^2 \bmod 2,$$

and that $t^2 + t + 1$ is irreducible mod 2. Hence from Condition 2, $\lambda = 1$, and the quotient knot has polynomial $t^2 + t + 1 \pmod 2$. However, $A_K(t)$ is irreducible, so Condition 1 rules this out.

EXERCISES

The exercises begin with a definition which will simplify notation.

☐ **DEFINITION.** *The total degree of a polynomial is the difference between the degrees of its highest and lowest degree nontrivial terms. The mod p total degree is the difference between the degrees of the highest and lowest degree terms having coefficients not divisible by p.*

3.1. Explain why the total degree of a product of two polynomials is the sum of their total degrees. Show this is also true mod p, for prime p.

3.2. Use the symmetry of the Alexander polynomial to show that the Alexander polynomial of a knot always has even total degree, integrally and mod p.

3.3. Apply Condition 2 to show that if a knot has prime power period $q = p^r$ then its polynomial has total mod p degree $2kq + (q-1)(\lambda - 1)$, where k is a nonnegative integer and λ is relatively prime to p.

3.4. (a) Use Exercise 3.3 to show that if a knot has prime period p and the mod p total degree of its Alexander polynomial is 2, then $p = 3$ ($k = 0$, $\lambda = 2$) or $p = 2$ ($k = 0$, $\lambda = 3$).

   (b) Show that if a knot has prime period p and the mod p total degree of its Alexander polynomial is 4, then either $p = 5$ ($k = 0$, $\lambda = 2$) or $p = 2$ ($k = 1$, $\lambda = 1$ or $k = 0$, $\lambda = 5$). (Remember that λ and p are relatively prime.)

   (c) If a knot has prime period p and the mod p total degree of its Alexander polynomial is 6, what are the possibilities for p and the corresponding k and λ?

3.5. Show that if a knot has period 7 and Alexander polynomial of mod 7 total degree 6 then its Alexander polynomial is $t^6 - t^5 + t^4 - t^3 + t^2 - t + 1$ (mod 7).

3.6. Show that no knot of 8 or fewer crossings has prime period 11 or more.

3.7. Show that the only knot with fewer than 9 crossings of period 7 is 7_1.

3.8. Show that if a knot has period 5 and Alexander polynomial of mod 5 total degree 4 then its Alexander polynomial is $t^4 - t^3 + t^2 - t + 1$ (mod 5). Use this to show that the only knot with fewer than 9 crossings of period 5 is 5_1.

3.9. Show that if the Alexander polynomial of a knot is $3t^2 - 5t + 3$, then the Murasugi conditions do not rule out period 3. State a general result encompassing this example.

3.10. Show that if the Alexander polynomial of a knot factors as the product of two irreducible cubics, and equals $(t+1)^6$ (mod 3) it cannot have period 3.

4 Periodic Seifert Surfaces and Edmonds' Theorem

If Seifert's algorithm for constructing Seifert surfaces for a knot is applied to the periodic diagram of a knot, the resulting surface displays the same periodic symmetry as the knot. Rather than call such a surface periodic, it is usually called *equivariant* . In general, a knot is called periodic if it can be deformed in such a way that it is fixed by the rotation R_q; a Seifert surface, F, is called equivariant, of period q, if it can be deformed so that $R_q(F) = F$. Hence, Seifert's algorithm implies that every period q knot bounds an equivariant Seifert surface of period q.

Equivariant Seifert surfaces become a useful tool with the use of the following theorem of Edmonds.

□ **THEOREM 3.**   *If a knot K is of period q, then there exists a period q equivariant Seifert surface, F, for K, with* genus$(F) = g(K)$.

As was noted earlier, Seifert's algorithm applied to a knot diagram might not produce a minimal genus Seifert surface, and for periodic knots the algorithm is no more efficient. Initially there is no reason to expect that symmetries of knots would be so strongly reflected in the surfaces that they bound.

The construction of a minimal genus equivariant Seifert surface involves a geometric technique not mentioned earlier, the use of minimal, or area minimizing, surfaces. If K is periodic, select a representative which is fixed by the rotation about the z-axis. Deep analytic results along with topological arguments imply that among all least genus Seifert surfaces for the knot there is one of least area. Edmonds proved that this area minimizing surface is equivariant.

THE RIEMANN–HURWITZ FORMULA AND THE PROPERTIES OF EQUIVARIANT SURFACES

In order to apply Theorem 3, the properties of equivariant surfaces must be developed. If K is periodic and has quotient knot J, then a Seifert surface for J can be lifted to give an equivariant Seifert surface for K. Conversely, any equivariant Seifert surface for K determines a Seifert surface for J. Figure 8.14 illustrates an equivariant view of the $(3,3,3,)$-pretzel knot, and its quotient knot. If Seifert's algorithm is applied to these diagrams, then the resulting surfaces are equivariant. The pretzel knot bounds a surface of genus 1, and the quotient is an unknot bounding a disk.

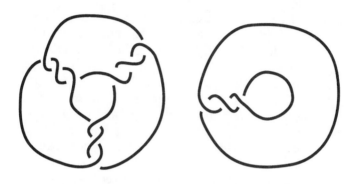

Figure 8.14

☐ **THEOREM 4.** (*Riemann–Hurwitz Formula*) *Let F be a genus g oriented surface which is equivariant with respect to a rotation about the z-axis of angle 360/q, and let G be the quotient of F. If both F and G have one boundary component, then*

$$\text{genus}(F) = q(\text{genus}(G)) + (q-1)(\Lambda - 1)/2,$$

where Λ is the number of points of intersection of F (or G) with the z-axis.

PROOF
The idea of the proof is fairly simple. Rather than compute the genus, compute the Euler characteristic. The Euler characteristic is defined in terms of a triangulation of the

surface, so to relate the two Euler characteristics one starts with an "equivariant" triangulation.

A triangulation of G can be picked so that the intersection points of G with the z-axis are all vertices; for this triangulation there is a corresponding triangulation of F. Every triangle in the triangulation of G determines q triangles in the triangulation of F. The same is true for the edges. All the vertices of G which are not on the z-axis lift to q vertices in F. The Λ vertices on the z-axis each lift to a single vertex in the triangulation of F.

Denoting the number of triangles, edges, and vertices in F by t_F, e_F and v_F, respectively, and using similar notation for G, one has $t_F = qt_G$, $e_F = qe_G$, and $v_F = qv_G - (q-1)\Lambda$. The argument is completed by computing the Euler characteristics in terms of the alternating sum of the number of triangles, edges, and vertices, and then algebraically translating into a formula for the genus. The details are left to the exercises.   □

To use these results, one final note about linking numbers is needed.

□ **THEOREM 5.** *If the linking number of a periodic diagram for K is λ then an equivariant Seifert surface for K intersects the z-axis in Λ points, where $\Lambda \geq \lambda$ and $\Lambda = \lambda \pmod 2$.*

PROOF
These relations follow easily from an alternative method of computing λ; using any Seifert surface, the linking number is given as the absolute value of the number of times the z-axis cuts the surface from its bottom minus the number of times it cuts it from the top.

The proof that this count gives λ is geometric, and is now sketched. The intersection of the right half of the

(x,z)-plane with a Seifert surface for K gives a family of arcs on the half plane. Some of the endpoints of the arcs are on the z-axis, and some are on the knot. The endpoints on the knot correspond to points of intersection of the positive x-axis in the knot diagram with the knot projection, and contribute to the count in the original definition of λ. Those on the z-axis correspond to points used in the count giving the alternative definition of λ just presented.

If an arc has both endpoints on the knot, then those two points will have opposite signs in the count giving λ, and cancel each other. Similarly for arcs with both endpoints on the z-axis. Arcs running from K to the z-axis give a pairing of the remaining points, and show that the two counts are equal. The remaining details consist of checking that signs work out as desired. $\qquad\square$

EXAMPLE
(Periods of genus 3 knots) The Riemann–Hurwitz formula places strong constraints on the possible periods of a surface, based on its genus. Consider, for example, a genus 3 surface, F, with one boundary component. Suppose that F is periodic of period q, and the quotient has genus g_G. Then the Riemann–Hurwitz formula implies that $3 = qg_G + (q-1)(\Lambda - 1)/2$

If $q > 3$, then g_G is clearly 0, and as a consequence $4 \leq q \leq 7$ and $2 \leq \Lambda \leq 7$. Checking cases shows the only possibilities are $q = 7$, $\Lambda = 2$, and $q = 4$, $\Lambda = 3$.

For $q = 3$ the equation becomes $3 = 3g_G + (\Lambda - 1)$, and the only solutions are $g_G = 1$, $\Lambda = 1$, and $g_G = 0$, $\Lambda = 4$. For $q = 2$ the equation becomes $3 = 2g_G + (\Lambda - 1)/2$, and in this case the only solutions are $g_G = 1$, $\Lambda = 3$, and $g_G = 0$, $\Lambda = 7$.

Applying Theorem 3, one has that the only possible periods of a genus 3 knot are 2, 3, 4, and 7.

As this example illustrates, Theorem 3 along with 4 and 5 yield strong relationships between the genus of a knot and its possible periods. In general these are captured by the following corollary.

□ **COROLLARY 6.** (*Edmonds' Conditions*) *If K is a periodic knot of period q, then there are nonnegative integers g_G and Λ such that $g(K) = qg_G + (q-1)(\Lambda-1)/2$. If a periodic representative of K has linking number λ with the z-axis, then $\Lambda \geq \lambda$ and $\Lambda = \lambda \pmod 2$, and λ is relatively prime to q.*

EXERCISES

4.1. Complete the proof of the Riemann–Hurwitz formula by computing the Euler characteristic of F in terms of that of G, and express the result in terms of the genus.

4.2. (a) Find all possible periods of a genus 1 surface with one boundary component. For each period what are the possible values of g_G and Λ?

    (b) Repeat the calculation for surfaces of genus 2 and 4.

4.3. Find an upper bound on the period of a knot based on its genus. Show that there are only a finite number of possibilities for the genus of the quotient knot and Λ.

4.4. Prove a converse to Theorems 4 and 5. That is, show that given nonnegative integers g_F, g_G, q, Λ, and λ, satisfying $g_F = qg_G + (q-1)(\Lambda-1)/2$, with $\Lambda \geq \lambda$, $\Lambda = \lambda \pmod 2$ and λ relatively prime to q, there is a period q equivariant surface of genus g_F with quotient of genus g_G. F should also intersect the z-axis Λ times, and its boundary should link the z-axis λ times.

4.5. For what values of g is there no surface of genus g and period 7? In general, show that for a given period q there are only a finite number of values for g such that there is no genus g surface of period q.

4.6. The $(3,3,3)$-pretzel knot, 9_{35}, has Alexander polynomial $7t^2 - 13t + 7$. Show that Murasugi's theorem does not eliminate the possibility of this knot having period 7, but that Edmonds' theorem does. Find further examples of pretzel knots which have periods ruled out by Theorem 3 but not by Theorem 2. Conversely, find examples for which Murasugi's criteria place constraints on the possible periods which cannot be obtained by genus considerations.

4.7. (a) Let K be of crossing index n with n odd. Exercise 4.4 of Chapter 7 states that if K is not a $(2,n)$-torus knot, then the genus of K is at most $(n-3)/2$. Using this, apply Edmonds' condition to find a bound on the possible periods of n crossing knots with n odd. (Answer: $q \leq n-2$.)

(b) If the crossing index of K is an even integer n, then Exercise 4.4 of Chapter 7 states that the genus of K is at most $(n-2)/2$. Find a bound on the periods of n crossing knots with n even. (Answer: $q \leq n-1$.)

5 Applications of the Murasugi and Edmonds Conditions

It is easily seen that there are instances where one of Theorems 2 or 3 can be applied to rule out a possible period for a knot and the other theorem does not apply. What is much more interesting is that there are examples of knots for which neither result alone places constraints on its period, but that when applied together limitations do occur. The interplay is provided by the quantity λ, which appears in both the Murasugi and Edmonds conditions. In this section that interplay will be demonstrated.

PERIOD 3 KNOTS

☐ **COROLLARY 7.** *If a genus* 1 *knot* K *has period* 3, *then its Alexander polynomial satisfies* $A_K(t) = \pm t^i(t^2 + 2t + 1)$ (mod 3).

PROOF
The only solution of the Edmonds condition, with $g(K) = 1$ and $q = 3$, is given by $g_G = 0$ and $\Lambda = 2$. (See Exercise 4.2 of the previous section.) Since the λ in Theorem 2 satisfies $\lambda \leq \Lambda$ with equality mod 2, and λ is relatively prime to 3, it follows that $\lambda = 2$. The result now follows from Theorem 2, since the degree of the Alexander polynomial of a genus 1 knot is at most 2. (One could also argue at this point that the quotient knot has genus 0 and hence trivial Alexander polynomial.)   ☐

☐ **COROLLARY 8.** *If a genus* 2 *knot* K *has period* 3, *then its Alexander polynomial satisfies* $A_K(t) = \pm t^i$ (mod 3).

PROOF
By Exercise 4.2 of the previous section, the only solution to the Edmonds condition with $g(K) = 2$ and $q = 3$ is given by $g_G = 0$ and $\Lambda = 3$. Hence, the quotient knot is trivial, and has trivial Alexander polynomial. It follows that λ in Murasugi's conditions is 1 or 3, but 3 is not possible, as λ and the period are relatively prime. The result follows.   ☐

EXAMPLES

The $(1,-3,5)$-pretzel knot (7_2 in the appendix) is a genus 1 knot with Alexander polynomial $-3t^2 + 7t - 3$. By Corollary 7 it cannot have period 3. This result does not follow from either Theorem 2 or Corollary 6 individually.

The knot in Figure 8.15 has Alexander polynomial $3t^4 - 7t^3 + 7t^2 - 7t + 3$. Thus, its genus is at least 2. Seifert's algorithm, which was introduced in Chapter 4, produces a genus 2 Seifert surface, so that the knot has genus exactly 2. As a consequence of this, Corollary 8 now implies that the knot does not have period 3. Again, this result does not follow from either Theorem 2 or Corollary 6.

Figure 8.15

PERIOD 5 KNOTS

□ **COROLLARY 9.** *If a nontrivial knot K is of period 5 and $g(K) \leq 3$, then the Alexander polynomial of K satisfies $A_K(t) = \pm t^i(t^4 - t^3 + t^2 - t + 1)$ (mod 5), and genus$(K) = 2$.*

PROOF

The only solution to the Edmonds condition with $q = 5$ and $g(K) \leq 3$ is given by $g(K) = 2$, $g_G = 0$ and $\Lambda = 2$. Since λ cannot be 0 it follows that $\lambda = 2$. The quotient knot is trivial, as it bounds a genus 0 Seifert surface, and hence has trivial Alexander polynomial. The result now follows from the Murasugi conditions.   □

Figure 8.16

EXAMPLE

The knot in Figure 8.16 has Alexander polynomial $5t^4 - 15t^3 + 21t^2 - 15t + 5$. Neither the Murasugi nor the Edmonds conditions individually rule out period 5. However, Corollary 9 applies, because Seifert's algorithm produces a genus 2 surface. (The knot is in fact genus 2, but this observation is not needed in order to apply Corollary 9.)

EXERCISES

5.1. The knot illustrated in Figure 8.17 has Alexander polynomial $6t^2 - 13t + 6$. Show that it does not have period 3.

Figure 8.17

5.2. Find all possible Alexander polynomials mod 7 for period 7 knots K, with $g(K) \leq 5$.

5.3. Describe all possible Alexander polynomials mod 3 for period 3 knots k_1 with $g(k) = 3$.

5.4. Show that if a nontrivial period 7 knot has crossing index less than 14, then its Alexander polynomial is of the form $t^6 - t^5 + t^4 - t^3 + t^2 - t + 1 \mod 7$. The knot shown in Figure 8.18 has Alexander polynomial $7t^4 - 21t^3 + 29t^2 - 21t + 7$. This provides an example of a knot that can easily be shown to not have period 7 using a combination of methods. (In this case, does either the Murasugi or Edmonds criteria apply individually to rule out period 7?)

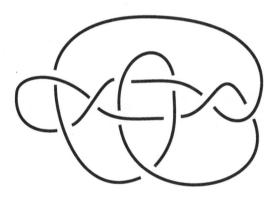

Figure 8.18

CHAPTER 9: HIGH-DIMENSIONAL KNOT THEORY

The theory of knots in R^3 naturally generalizes to a study of knotting in R^n, with $n > 3$, and many new and fascinating aspects of knot theory appear in this high-dimensional setting. What is perhaps most surprising is that many problems that are intractable in the classical case have been solved for high-dimensional knots. There is also a strong interplay between knot theory in different dimensions, and this interplay leads to an array of new topics at the border of the classical and high-dimensional settings.

The definitions of polygonal knot and of deformation of knots generalizes immediately to R^4, (or for that matter R^n); one can simply consider sequences of points in 4-space instead of R^3. Knots formed in this way are called 1-dimensional knots in 4-space, or, more briefly, 1-knots. It turns out though that there is really no interesting theory of such 1-knots; all such knots in 4-space are equivalent.

The appropriate generalization increases the dimension of the knot as well as the dimension of the ambient space. The definition of surface given in Chapter 4 easily generalizes to yield a definition of surfaces in 4-space. A 2-knot is a surface in R^4 that is homeomorphic to S^2, the standard sphere in 3-space. Figure 9.1 is a schematic illustration of such a knot. Section 1 discusses some of the details of the definitions, as well as a new subtlety that arises at the foundation of the subject.

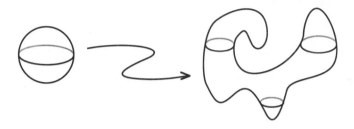

Figure 9.1

One of the pleasures of studying high-dimensional knot theory is the discovery that it is possible to visualize, and sketch, knots in higher dimensions. Sections 2 and 3 demonstrate this, and should help the reader gain intuition about R^4 and the knotting phenomena that occur there.

Section 4 describes a property of classical knots in R^3, called *sliceness*, that is defined in terms of knotted 2-spheres in R^4. It is here that the interplay between dimensions 3 and 4 comes out. The notion of slice knot can be used to define an equivalence relation called *concordance* on the set of classical knots, and it turns out that there is a natural (abelian) group structure on the set of concordance classes of knots. Very little is known about this group, and what is known is summarized in Section 5.

1 Defining Knots in Higher Dimensions

The definition of *knot* given in Chapter 2 ruled out the possibility of infinite knotting, as illustrated in Figure 2.2. In higher dimensions the definitions must also rule out such pathologies. As in the classical case, this can be done using either polygonal knots or smooth knots. Unlike the classical case, the two theories that arise can be distinct. For instance, it is true that every smooth knot can be closely approximated by a polygonal knot, and that if two different approximations are chosen, the two polygonal knots are equivalent. However, when two inequivalent smooth knots are approximated by polygonal knots, it is possible that the resulting knots may be polygonally equivalent.

The study of the distinctions between the theories lies at the foundations of topology and is beyond the scope of this chapter. Only the smooth theory, probably the easiest to describe, is summarized. The discussion begins with the definition of the k-sphere.

☐ **DEFINITION.** *The k-sphere, S^k, is the set of unit vectors in R^{k+1}; that is,*

$$S^k = \{(x_1,\dots,x_{k+1}) \in R^{k+1} \mid x_1^2 + \cdots + x_{k+1}^2 = 1\}.$$

Two important examples are S^1, which is the unit circle in the plane, and S^2, which is the 2-sphere illustrated in Figure 9.1. (The convention of calling these spheres k-spheres rather than $(k+1)$-spheres is based on the fact that intrinsically S^k is k-dimensional, it is only a subset of a $(k+1)$-dimensional space.)

☐ **DEFINITION.** *A smooth knotted k-sphere in R^n, K, is a subset of R^n of the form $F(S^k)$, where F is a one-to-one differentiable function from S^k to R^n with everywhere nonsingular derivative.*

(Recall that the derivative of such a function assigns to each point p in S^k a linear map, $D_p(F)$, from the set of tangent vectors to S^k at p to R^n. The linear map is nonsingular if its image is k-dimensional. Also, note that this definition corresponds to the definition of smooth knot given in Chapter 1 in the case of $k = 1$ and $n = 3$. The nonsingularity condition eliminates the type of knotting that was illustrated in the second figure of Chapter 2.)

The definition of equivalence is harder. Suppose that K_0 and K_1 are smooth k-knots in R^n. They are considered smoothly equivalent if there is a family of differentiable functions, F_t, $0 \le t \le 1$, from S^k to R^n such that:

(1) for all t, F_t is one-to-one with everywhere nonsingular derivative,

(2) $F_0(S^k) = K_0$ and $F_1(S^k) = K_1$, and

(3) the function G from $S^k \times [0,1]$ to R^n defined by $G(p,t) = F_t(p)$ is differentiable.

Roughly stated, two knots are equivalent if one can be smoothly deformed into the other through a sequence of smooth knots.

EXERCISES

1.1. Describe the 0-sphere, S^0. Explain why all 0-knots in R^2 are equivalent.

1.2. (a) Give a definition of a high-dimensional link. Your definition should include the possibility of having components of different dimension.

　　 (b) Define equivalence of high-dimensional links.

2 Three Dimensions from a 2-dimensional Perspective

In Edwin A. Abbott's novel *Flatland* the narrator tries to describe a 2-sphere to the inhabitants of a plane. As Abbot intended, that description leads to an understanding of how 4-dimensional phenomena can be studied from a 3-dimensional perspective.

When a plane parallel to the x–y plane is lowered through space, its intersections with a 2-sphere give a series of 2-dimensional cross-sections of the sphere. The first non-trivial cross-section consists of a single point. The point then opens up into a circle which grows until its radius is that of the sphere, and then it shrinks down to a point and disappears. This sequence of cross-sections forms the frames of a Flatlandian movie of a sphere in 3-space.

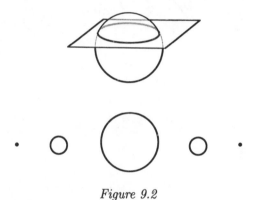

Figure 9.2

Flatlandian movies could be made that would illustrate other surfaces in 3-space. For instance, Figure 9.3 shows some frames from a description of a surface homeomorphic to a sphere and Figure 9.4 illustrates a Flatlandian film of a torus.

Knots could also be shown to Flatlanders as a series of 2-dimensional cross-sections. For instance, the first nontrivial cross-section of the trefoil might consist of two points in the plane. Each of those points immediately split into a pair of points. The center two points then rotate

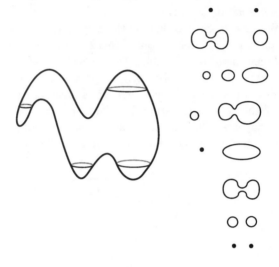

Figure 9.3

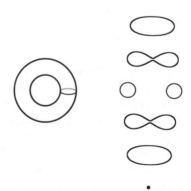

Figure 9.4

about one another. Finally, pairs of points rejoin and then disappear. This is in Figure 9.5.

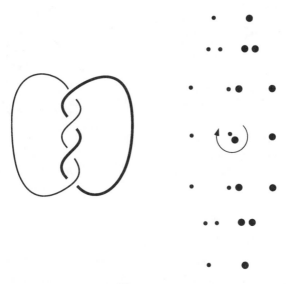

Figure 9.5

The exercises ask you to describe other knots and links to Flatlanders.

EXERCISES

2.1. Draw a series of frames illustrating cross-sections of the sphere illustrated in Figure 9.6.

2.2. Draw a series of figures illustrating the cross-sections of a pair of nested spheres.

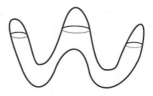

Figure 9.6

2.3. Find an algorithm for deciding if a surface described via a series of cross sections is connected.

2.4. Describe the cross sections of the Figure-8 knot and the Borromean rings.

2.5. The (p,q)-torus knot can be placed so that its cross-sections are extremely simple to describe. Find that description.

2.6. Draw 2-dimensional cross-sections of the knotted torus illustrated in Figure 9.7.

Figure 9.7

3 Three-dimensional Cross-sections of a 4-dimensional Knot

Just as 3-space can be swept out by a plane, 4-space can also be swept out by 3-dimensional hyperplanes. Often the fourth coordinate is viewed as the time, or t, coordinate and the hyperplanes are viewed as parameterized by time. To be precise, let $H_\tau = \{(x,y,z,t) \in R^4 \mid t = \tau\}$. A 2-knot in 4 space, K, can be described via a sequence of 3-dimensional cross-sections of the form $H_t \cap K$; many of these intersections might be classical knots or links in H_t, which is naturally identified with R^3. The simplest such sequence begins with a point in R^3 which immediately opens up into an unknotted circle. The circle grows for a while and then shrinks back into a point and disappears. This is similar to the sequence illustrated in Figure

9.2. Corresponding to Figure 9.3 there is another picture of a 2-sphere in 4-space which begins with 2 points. Both of the 2-knots described by these sequences are in fact trivial, where a trivial 2-knot in R^4 is a knot which can be deformed into the standard S^2 in $H_0 = R^3$.

Of much greater interest is the sequence of cross-sections drawn in Figure 9.8. Here only the first half of the series is illustrated.

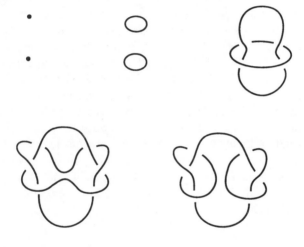

Figure 9.8

The second half appears as the first in reverse order. The 2-knot that this describes is called the *spun trefoil*, and was among the first examples of nontrivial 2-knots. It was discovered by Artin.

Proving that the spun trefoil is nontrivial requires algebraic topology, and in particular a careful study of its fundamental group. There actually is a method of showing that it is knotted which is based on colorings, but a proof of

the validity of this approach depends on a careful study of
the fundamental group in any case. Unfortunately, there is
no simple generalization of Reidemeister moves in dimension 4. A description of colorings will be given later in the
section.

Other examples of 2-knots are easily constructed in
a similar manner. The exercises ask you to form a few.
An interesting family of examples, described by Zeeman,
can be formed by slightly modifying the sequence of cross-
sections just described. The first half of the sequence is
the same as for the spun trefoil. Now, before reversing
the sequence to construct the "bottom half" of the knot,
one of the two trefoils can first be twisted about its axis
as illustrated in Figure 9.9. If it is twisted k times, the
resulting knot is called the k-twist spin of the trefoil. One
of Zeeman's remarkable discoveries was that the 1-twist
spin of the trefoil (or any other knot for that matter) is
actually unknotted. An immediate consequence of this is
that the unknotted 2-sphere in 4-space can be deformed
so that some of its cross-sections are nontrivial knots in
3-space! (Stallings first constructed an example of this
phenomena prior to Zeeman's work.)

Figure 9.9

One useful fact in studying knotted 2-spheres is that any 2-knot in R^4 can be slightly deformed so that all but a finite number of cross-sections are either classical links or empty. The finite set of exceptions correspond to transitions where components appear or disappear, or else components band together or split apart. Operations of this second type are called *band moves*. These two types of transitions are illustrated in Figure 9.10. The descriptions of 2-knots already presented had this property, and for the rest of the chapter all knots will be assumed to have such cross-sections.

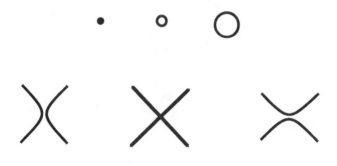

Figure 9.10

COLORING 2-KNOTS

A proof that if a classical knot or link diagram is colorable then every diagram for that link is colorable was given in Chapter 3, and was based on the Reidemeister moves. The proof actually showed more: once a coloring is picked for a diagram, then there is a *canonical* choice of coloring for all other diagrams. To see this, observe that when a Reidemeister move is performed on a colored diagram there is a natural choice of coloring for the new diagram.

A 2-knot is called colorable if every cross-section can be colored so that nearby cross-sections are colored in a consistent manner, and at least two colors appear. When new components appear there is no restriction on how they are colored, but when components join together, the colorings have to be the same at the point that they meet. A nontrivial coloring of the spun trefoil is illustrated in Figure 9.11.

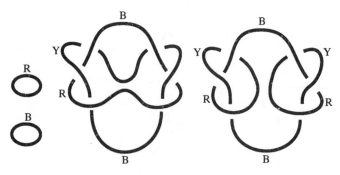

Figure 9.11

It is left to the reader to check that if the trivial components in the first frame are colored red and blue, then the coloring of the second frame must be as illustrated. More examples are presented in the exercises.

EXERCISES

3.1. Illustrate the cross-sections of a knotted 2-sphere for which the middle cross-section is as illustrated in Figure 9.12a. The dotted line provides a hint.

3.2. Repeat Exercise 1 for the knot illustrated in Figure 9.12b. Why is the surface you construct a 2-sphere?

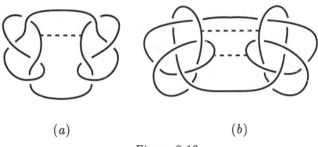

(a) (b)

Figure 9.12

3.3. Repeat Exercise 1 for the knots illustrated in Figure 9.13.

Figure 9.13

3.4. Find a nontrivial coloring or show that one does not exist for each of the knots constructed in the previous exercises.

3.5. For any knot K, the knot $K \# K^{rm}$, as illustrated in Figure 9.12, occurs as the middle cross-section of a knotted

2-sphere in R^4, called the *spin* of K. Describe the general construction of the 2-knot. Show that if K is colorable, then so is the spin of K.

3.6. The sequence of drawings in Figure 9.14 illustrate a surface in 4-space. The surface is not a sphere. What is it?

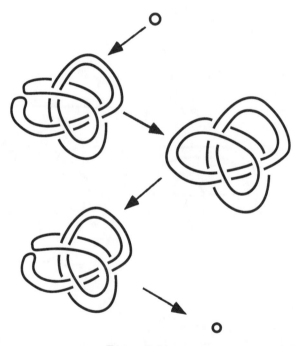

Figure 9.14

3.7. The sequence of diagrams in Figure 9.15 (taken from [Fox]) illustrate the cross-sections of a knotted surface in 4-space for which one cross-section is a nontrivial link. Verify that the surface is a 2-sphere. (Epstein showed that this 2-knot is in fact trivial.)

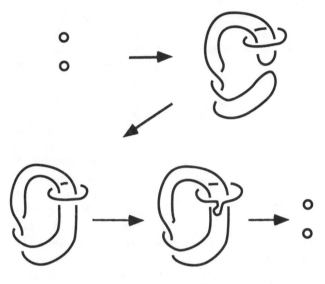

Figure 9.15

3.8. The theory of mod(p) labelings applies to 2-knots also. Use this to show that the knots constructed in Exercise 1 are nontrivial.

4 Slice Knots   The last section presented several examples of classical knots in R^3 that arise as cross-sections (or slices) of 2-knots in R^4; these were illustrated in Figures 9.8, 9.12 and 9.13. Such knots are called *slice knots*. Determining if a knot is slice is a fascinating problem at the border between classical knot theory and the high-dimensional theory.

There is a useful alternative definition of slice knot. Note that the x–y plane in 3-space is the boundary of upper half space, $R_+^3 = \{(x,y,z) \mid z \geq 0\}$. Similarly, 3-space, R^3, can be identified with the x–y–z hyperplane (H_0) in upper 4-space, $R_+^4 = \{(x,y,z,t) \mid t \geq 0\}$. A classical knot in R^3 is slice if it is the boundary of a *smooth disk* in R_+^4, and the disk it bounds is called its *slice disk*. A slice disk is illustrated schematically in Figure 9.16.

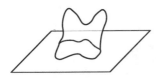

Figure 9.16

The equivalence of the two definitions of slice knot is fairly simple to describe. If a knot is the cross-section of a knotted 2-sphere, K, then the portion of K lying above the cross-section forms a slice disk. On the other hand, if a classical knot bounds a slice disk, then the union of the disk and the mirror image of that disk in lower 4-space, R_-^4, forms a knotted 2-sphere.

It was seen in the previous section, Exercise 3.5, that the connected sum of a knot and its mirror image is always

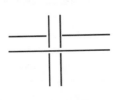

Figure 9.17

slice. (Actually, some care with orientation is needed here; the correct statement is that $K \# K^{rm}$ is slice.) After describing a few more examples, methods of proving that a knot cannot be sliced will be described. A knot is called a *ribbon knot* if it bounds a disk with self intersections only of the type illustrated in Figure 9.17. Such a disk is called a *ribbon disk.*

Two examples of ribbon knots were drawn in Figure 9.13, the first is actually the square knot, which was the first example of a slice knot in the previous section. The diagrams make it clear how the name ribbon arises.

□ **THEOREM 1.** *Every ribbon knot is slice.*

PROOF
A series of cross-sections of a slice disk is easy to construct. Near the ribbon intersections of the ribbon disk the knot can be pinched to divide it up into several components. That collection of components forms an unlink and each one can be shrunk to a point. This is illustrated in Figure 9.18 below. Usually this algorithm is excessive in that a slice disk could be found without introducing so many components.   □

Figure 9.18

□ **CONJECTURE.** *Every slice knot is ribbon.*

This conjecture has been known since the early 1960's and remains one of the most challenging problems in the field.

CONDITIONS ON SLICE KNOTS

If a knot is slice there are strong restrictions on its possible Seifert matrices. These in turn place restrictions on the Alexander polynomial and signatures of slice knots. The proof of the main theorem is beyond what can be presented here, but the corollaries follow fairly easily.

□ **THEOREM 2.**   *If a knot K is slice and V is any Seifert matrix arising from a Seifert surface of genus g, then there is an invertible (determinant 1) integer matrix M such that MVM^t is of the form*

$$\begin{pmatrix} 0 & B \\ C & D \end{pmatrix}$$

where B, C, and D are $g \times g$ matrices with $B - C = \pm I_g$, where I_g is the $g \times g$ identity matrix.

□ **COROLLARY 3.**   *The Alexander polynomial of a slice knot can be factored as $\pm t^k f(t) f(t^{-1})$ for some integer polynomial f and integer k.*

PROOF

The Alexander polynomial is given by

$$\det(V - tV^t) = \det(M(V - tV^t)M^t)$$
$$= \det(MVM^t - tMVM^t).$$

This last matrix is of the form

$$\begin{pmatrix} 0 & B - tC^t \\ C - tB^t & D - tD^t \end{pmatrix}$$

and $f(t)$ can be taken to be $\det(B - tC^t)$.   □

EXAMPLE

The trefoil knot is not slice, as its Alexander polynomial is irreducible.

This result is quite useful, but it fails for a knot as simple as the granny, the connected sum of the trefoil with itself.

□ **COROLLARY 4.** *If a knot is slice then its signature (and all its ω-signatures) are 0.*

PROOF

Only the real signature will be discussed; a proof for the complex signatures is similar.

The matrix $V + V^t$ which must be diagonalized can be put in the form

$$\begin{pmatrix} 0 & S \\ S^t & R \end{pmatrix}$$

and since $V + V^t$ is invertible over the reals, the matrix S is invertible. Hence, simultaneous row and column operations can be used to put this matrix into the form

$$\begin{pmatrix} 0 & I_g \\ I_g & R \end{pmatrix}.$$

Further row and column operations can be used to eliminate the bottom right-hand block. Finally, it is a straightforward calculation that the signature of the matrix

$$\begin{pmatrix} 0 & I_g \\ I_g & 0 \end{pmatrix}$$

is 0.   □

EXAMPLE

As the signature of the trefoil is 2, the granny has signature 4, and is not slice.

If a knot has a Seifert form which is similar to a matrix of the form

$$\begin{pmatrix} 0 & B \\ C & D \end{pmatrix}$$

as above, then it is called *algebraically slice*. Theorem 2 thus states that every slice knot is algebraically slice. In higher dimensions there are corresponding notions of slice knots and algebraically slice, and results of Kervaire and Levine imply that in higher dimensions a converse result holds; a high-dimensional knot is slice if and only if it is algebraically slice. The surgery theoretic methods used in their proofs fail for classical knots, and Casson and Gordon proved that there are algebraically slice knots in R^3 that are not slice. This is described further in Exercise 4.5.

EXERCISES

4.1.   Prove that if a knot can be reduced to an unlink of $n+1$ components by performing n band moves, then it is a ribbon knot. (It should be clear that it is slice.) Hence, the solution of Exercise 2.5 shows that knots of the form $K \# K^{rm}$ are actually ribbon.

Figure 9.19

4.2.   Show that the knot in Figure 9.19 is a ribbon knot. Also, argue that if the knot has n twists instead of 4 as illustrated, then for n odd it is genus 1 and hence is a prime slice knot. Finally show that it is the cross-section of a colorable 2-knot for all n.

4.3. Which knots with 7 or fewer crossings have polynomials satisfying the polynomial condition on slice knots given by Corollary 3.

4.4. The n-twisted double of the unknot, illustrated in Figure 9.20, with $n = 2$ has Seifert matrix V given by

$$\begin{pmatrix} -1 & 1 \\ 0 & n \end{pmatrix}.$$

For what values of n does the polynomial satisfy the conditions of Corollary 3? Show that for these values of n the

Figure 9.20

knot is actually algebraically slice. (Hint: If a quadratic polynomial factors as desired, it has rational roots.)

4.5. The doubled knots of the previous exercise are slice only when $n = 0$ or 2, as proved by Casson and Gordon. Show that for $n = 2$ the knot is ribbon. (Hint: consider the examples in Exercise 4.2.)

4.6. Not every unknotting number 1 knot is slice, as was seen in the previous exercises. However, every unknotting number 1 knot is the boundary of a genus 1 surface in R_+^4. Show this by finding a pair of band moves that changes an unknotting number 1 knot into a two component link and then into the unknot. The corresponding surface, which is completed by letting the unknot shrink to a point, is of genus 1. Why? (The bands can all be drawn near the crossing that needs to be changed.)

4.7. A ribbon knot is always the slice of a 2-knot, as described in the proof of Theorem 1. Show that if that 2-knot is colorable, then at the middle cross-section parallel arcs of the ribbon in the diagram are colored the same color.

4.8. (a) A link of two components in R^4 is called *splittable* if it can be deformed so that the components lie on opposite sides of the (y, z, t)-hyperplane. Show that the number of colorings of a split link, including trivial colorings, is the product of the number of colorings for each component.

(b) Figure 9.21a illustrates a link of two components which is the cross-section of a link of two 2-knots in R^4. Show that it is not splittable by counting colorings.

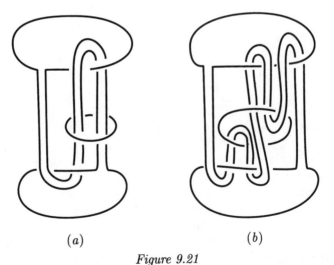

(a)   (b)

Figure 9.21

(c) Repeat part b for Figure 9.21b. This example is more interesting, as each component taken individually is trivial. (Can you see why?)

5 The Knot Concordance Group

Using the notion of sliceness, an equivalence relation called *concordance* can be placed on the set of classical knots. The operation of connected sum induces a group structure on the set of concordance classes of knots, and understanding the structure of this group is one of the outstanding problems in knot theory.

□ **DEFINITION.** *Knots K and J are called concordant if $K \# J^{rm}$ is slice.*

□ **THEOREM 5.** *Concordance forms an equivalence relation on the set of knots.*

PROOF

That the relation is reflexive follows from the earlier observation that $K \# K^{rm}$ is always slice. That it is symmetric is automatically satisfied. Transitivity is quickly reduced to showing that if knots K and $K \# J$ are slice, then so is J; the proof of this statement calls for a geometric construction which, although not too difficult, is not presented here.   □

□ **LEMMA 6.** *If K_1 is concordant to K_2 and J_1 is concordant to J_2, then $K_1 \# J_1$ is concordant to $K_2 \# J_2$.*

PROOF

The proof calls for a geometric fact which has not been proved in the text; if knots K and J are slice, then so is $K \# J$. The reader should be able to sketch the argument. The rest of the proof is formal. One needs to show that

$$(K_1 \# J_1) \# (K_2 \# J_2)^{rm}$$

is slice. However, this knot is the same as

$$(K_1 \# K_2^{rm}) \# (J_1 \# J_2^{rm}),$$

which is the connected sum of two slice knots.   □

This lemma implies that the connected sum operation induces a well-defined operation on the set of concordance classes of knots.

□ **THEOREM 7.**   *With respect to the operation induced by connected sum, the set of concordance classes of knots forms an abelian group, denoted C_1^3.*

PROOF

Associativity follows from the fact that connected sum of knots is an associative operation. Similarly for commutativity. The identity element is given by the concordance class of the unknot, U, since $K \# U = K$. (The concordance class of the unknot consists of all slice knots.) The inverse of the concordance class of K is given by the concordance class of K^{rm}, since $K \# K^{rm}$ is slice.   □

THE STRUCTURE OF C_1^3

As mentioned earlier, understanding the structure of

this group is one of the outstanding problems of low-dimensional topology. The few facts that are known about the group can be easily summarized here.

Fact 1: The concordance group is countable.

Every knot can be deformed so that all its vertices have rational coordinates. It follows that there are only a countable number of knot types, so certainly there are only a countable number of concordance classes.

Fact 2: The function that sends K to $\sigma(K)/2$ is a homomorphism from the concordance group onto Z, and hence the concordance group is infinite. (Here $\sigma(K)$ is the signature function defined in Chapter 6.)

That $\sigma(K)/2$ is a homomorphism follows from the observation that signature adds under connected sum (Exercise 3.6 of Chapter 6) and Corollary 4. (It is easily seen that $\sigma(K)$ is always even; see Exercise 3.4 of Chapter 6.) The trefoil has signature 2, and surjectivity follows.

Fact 3: There are elements of order 2 in the concordance group.

The figure-8 knot, K, provides one such example. As its Alexander polynomial is irreducible, it is not slice. However, it is negative amphicheiral so that $K = K^{rm}$ and hence $K \# K$ is slice.

Fact 4: C_1^3 maps homomorphically onto Z^∞.

An infinite collection of homomorphisms to Z can be defined using ω-signatures, and these can be pieced together to define the desired homomorphism. Levine found examples demonstrating surjectivity.

Beyond these few observations little more is known; for instance, whether or not there are elements of finite order other than order 2 is unknown. Recent advances in low-dimensional topology only indicate that this problem is even more complicated than anticipated.

It is possible to define concordance groups of higher-dimensional knots, and surprisingly the structure of these groups is well understood. Letting C_n^{n+2} denote the concordance group of n-knots in R^{n+2}, it is known that C_n^{n+2} is trivial for n even, and is isomorphic to

$$(Z/2Z)^\infty \oplus (Z/4Z)^\infty \oplus (Z)^\infty$$

for n odd.

EXERCISES

5.1. Use the result that whenever the knots K and $K \# J$ are slice then J is also slice to prove that concordance is transitive.

5.2. Use Alexander polynomials to prove that the trefoil and the $(2,5)$-torus knots are not concordant.

5.3. The $(2,p)$-torus knot has Alexander polynomial $(t^p + 1)/(t+1)$, which is irreducible for p prime. Use this to prove that for p prime the $(2,p)$-torus knot is not concordant to a knot of genus less than $(p-1)/2$.

5.4. (Casson) Although, by Exercise 4.6, unknotting number 1 knots always bound genus 1 surfaces in R_+^4, there are unknotting number 1 knots that are not concordant to genus 1 knots. Find an example of this.

CHAPTER 10:
NEW COMBINATORIAL TECHNIQUES

New combinatorial knot invariants have been discovered which are simple in definition and yet extremely powerful. Unlike those described earlier, there is no known connection to knot theory in higher dimensions. It now seems likely that they relate to properties that are unique to dimension 3. The new techniques have their roots in an observation made by Alexander in his original paper on the Alexander polynomial, an observation that went unexploited for forty years.

Given an oriented link diagram, focus on a particular crossing. If that crossing is changed from right to left or vice versa, a new link diagram results. The crossing could also be *smoothed* to obtain yet another link diagram. The smoothing process removes small sections of the arcs that pass over and under, and replaces them with a new pair of arcs that do not cross. There is only one way of doing this while maintaining the orientation of the original diagram. Hence, for a given diagram and crossing, there are a total of three associated diagrams, corresponding to links denoted L_+, L_-, and L_s. This is illustrated in Figure 10.1. Of course, the links that result depend on the choice of crossing.

Alexander proved that if his algorithm for computing the Alexander polynomial is applied *appropriately* to all three diagrams, the resulting polynomials are related by the equation $A_{L_+}(t) - A_{L_-}(t) = (1-t)A_{L_s}(t)$. This result makes it appear that the polynomials of L_- and L_s determine that of L_+. Unfortunately, this does not follow; the Alexander polynomial is only defined up to multiples of $\pm t^k$, and different choices of representative polynomials lead to different sums. However, as will soon be seen, there is a way to normalize the Alexander polynomial that makes this problem disappear.

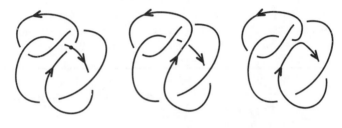

Figure 10.1

(A definition of the Alexander polynomial of oriented links was not presented in the text. The combinatorial approach of Chapter 3 extends to links, as does the definition in terms of Seifert matrices, $A_L(t) = \det(V - tV^t)$, where V is the Seifert matrix arising from an oriented Seifert surface for L.)

EXERCISES

0.1. It seems that by picking the appropriate crossing that is changed or smoothed in a given link diagram, both of the resulting links are somehow simpler. This problem asks you to formalize this.

Recall first that any link diagram can be changed into a diagram for the unlink by changing some of the crossings.

Define the *complexity* of a link diagram, D, to be the ordered pair of nonnegative integers $\chi(D) = (c^*(D), u^*(D))$, where $c^*(D)$ is the number of crossings in D and $u^*(D)$ is the minimum number of crossings in the link diagram that can be changed to create an unlink. Order the set of complexities by the rule $(c_1^*, u_1^*) < (c_2^*, u_2^*)$ if: (1) $c_1^* < c_2^*$ or (2) $c_1^* = c_2^*$ and $u_1^* < u_2^*$. (This ordering is called *lexicographical*, and is sometimes referred to as the *dictionary order*.)

(a) For a given diagram, D, if $u^*(D) \neq 0$, show that for some choice of crossing, changing the crossing and smoothing the crossing both result in diagrams of smaller complexity.

(b) Show that there is no infinite sequence of decreasing complexities, $\chi_1 > \chi_2 > \chi_3 > \ldots$.

1 The Conway Polynomial of a Knot

The Alexander polynomial of a knot, K, can be normalized so that it is symmetric, in the sense that $A_K(t) = A_K(t^{-1})$, and $A(1) = 1$. For example, the trefoil knot has polynomial $t - 1 + t^{-1}$. This symmetry is discussed in general in Exercise 1.2. Once normalized in this way, it can be written as a polynomial of the form $\nabla_K(z)$, where $z = t^{1/2} - t^{-1/2}$, and only positive powers of z appear in $\nabla_K(z)$. This new polynomial is called the *Conway polynomial*, or *potential function*, of K, $\nabla_K(z)$.

A simple calculation demonstrates this change of variable in the case of the trefoil polynomial; $t - 1 + t^{-1} =$

$z^2 + 1$. As a more complicated illustration, check that

$$2t^3 - 7t^2 + 13t - 15 + 13t^{-1} - 7t^{-2} + 2t^{-3}$$
$$= 2z^6 + 5z^4 + 3z^2 + 1,$$

again with $z = t^{1/2} - t^{-1/2}$. The exercises ask you to compute more examples and to prove the general result showing that every symmetric polynomial can be written in terms of z.

(The Alexander polynomial for links display the same symmetry, with one slight subtlety; one needs to consider the Alexander polynomial as a polynomial in the variable $t^{1/2}$, and it is well defined only up to multiples of $\pm(t^{1/2})^k$. This technical detail is discussed in Exercise 1.2. In any case, with a little care the Conway polynomial can be defined for links as well as for knots.)

Conway proved that the potential functions of links L_+, L_-, and L_s which are related as above, satisfy the recursion relation

$$\nabla_{L_+}(z) - \nabla_{L_-}(z) = -z\nabla_{L_s}(z).$$

This relation, along with the fact that for the unknot U, $\nabla_U(z) = 1$, leads to an efficient means for computing $\nabla_L(z)$.

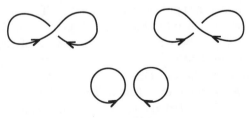

Figure 10.2

EXAMPLES

In Figure 10.2, L_+ and L_- are unknots, and L_s is an unlink of two components. It follows that for the unlink of two components $U_2, \nabla_{U_2}(z) = 0$. A similar argument shows that for *any* splittable link the Conway polynomial is trivial. (A link is called *splittable* if it can be deformed so that one component is on one side of the (y,z)-plane and the other components lie on the other side.)

In Figure 10.3, L_+ is the $(2,2)$-torus link, T_2, L_- is the unlink, and L_s is the unknot. It follows from the recursion relation that $\nabla_{T_2}(z) = -z$. If the orientation of one of the components is changed, the resulting Conway polynomial is $\nabla_L = z$.

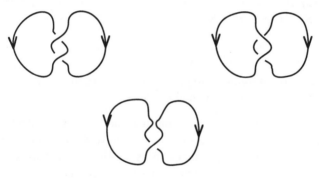

Figure 10.3

One can proceed to build up collections of examples in this way. Figure 10.4 relates the trefoil to the unknot and the $(2,2)$-torus link discussed above, and the recursion relation then shows that the Conway polynomial of the trefoil is $z^2 + 1$. The exercises ask you to consider a few more examples, and, in particular, to show that the $(2,5)$-torus knot has Conway polynomial $z^4 + 3z^2 + 1$.

The following theorem offers computational tools that are useful in more complicated examples.

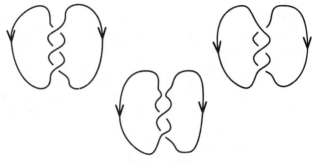

Figure 10.4

□ **THEOREM 1.**

 (a) *For knots K_1 and K_2, $\nabla_{K_1 \# K_2}(z) = \nabla_{K_1}(z)\nabla_{K_2}(z)$.*

 (b) *For any knot K, $\nabla_K(z) = \nabla_{K^m}(z) = \nabla_{K^r}(z)$.*

PROOF
All three statements follow from similar results concerning the Alexander polynomial of knots.   □

As a final example, consider the knot K illustrated in Figure 10.5. In the illustration it is shown how a sequence of crossing changes and smoothings can reduce K to simple knots and links, each of which has easily computed Conway polynomial. Applying the recursion formula yields $\nabla_K(z) = z^6 + 5z^4 + 4z^2 + 1$. In the figure, simplifications of the diagrams have been carried out that should be checked by the reader.

RECURSIVE DEFINITION OF THE CONWAY POLYNOMIAL
The recursive formula for the Conway polynomial, along with the condition that $\nabla_U(z) = 1$, offers an effective means for computing its value. These two conditions also offer a new means of defining the invariant.

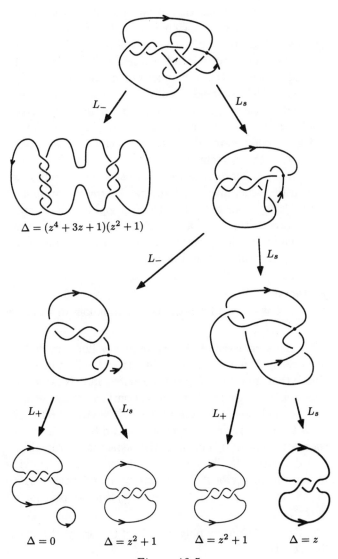

Figure 10.5

As has been seen with a few examples, given an oriented link L, a series of choices, both of diagrams and crossings, leads to a calculation of the value of $\nabla_L(z)$. One proof that the outcome is independent of the choices is that it is equivalent to the Alexander polynomial, suitably normalized, which is already known to be well defined. As an alternative, there is a direct proof that none of the choices made affect the outcome. As can be imagined, the proof is a delicate combinatorial argument which includes a careful analysis of the Reidemeister moves.

At this point it might seem that such a direct approach is of little value, given that several alternative arguments exist. The importance is both philosophical and practical; it reveals that such a recursive formula offers a means of actually defining invariants, and indicates a means for proving that they are well defined.

EXERCISES

1.1. Express several Alexander polynomials in terms of $z = t^{1/2} - t^{-1/2}$.

1.2. The symmetry of the Alexander polynomial follows most easily from the definition in terms of Seifert matrices. Recall that from the fact that the Seifert matrix is $2g \times 2g$, where g is the genus of the Seifert surface, one concludes that the polynomial satisfies $t^{2g}A(t^{-1}) = A(t)$. From this, show that the Alexander polynomial can be normalized so that $A(t^{-1}) = A(t)$. For links the Seifert matrix might be odd dimensional. Show that in this case, by multiplying by an odd power of $t^{1/2}$ one arrives at a function, $A(t)$, satisfying $A(t^{-1}) = -A(t)$, where $A(t)$ is now a polynomial in $t^{1/2}$.

1.3. Find the Conway polynomials of the oriented links in Figure 10.6.

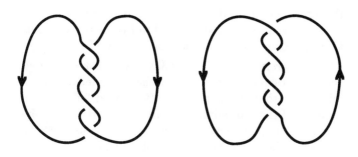

Figure 10.6

1.4. Show that the Conway polynomial of the $(2,5)$-torus knot is $z^4 + 3z^2 + 1$.

1.5. Compute the Conway polynomial for several knots in the appendix. Check each result by comparing it to the Alexander polynomial.

1.6. Give a "recursive" proof that the Conway polynomial of the connected sum of links, $L_1 \# L_2$, as illustrated in Figure 10.7, is the product of their Conway polynomials.

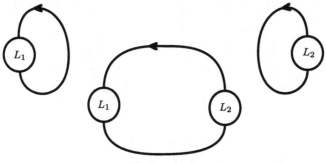

Figure 10.7

(Hint: A repeated sequence of crossing changes and deformations applied to L_1 express $\nabla_{L_1}(z)$ as a sum of terms of the form $z^{a_i} \nabla_{J_i}(z)$, where J_i is either an unknot or unlink. The same sequence can be applied to $L_1 \# L_2$ to express its Conway polynomial as the sum of terms $z^{a_i} \nabla_{J_i \# K_2}(z)$.)

1.7. In Exercise 5.5 of Chapter 3, you were asked to prove that a particular 11-crossing knot had Alexander polynomial 1. At that time, the calculation called for the computation of the determinant of a 10×10 matrix with polynomial entries, a daunting task without the assistance of a computer. Compute its Conway polynomial.

1.8. Suppose that a circle in the plane intersects a knot diagram in exactly four points, as illustrated for a particular example in Figure 10.8. Rotating that portion of the diagram in the circle by 180 degrees creates a new link diagram. The new link is called a *mutant* of the first.

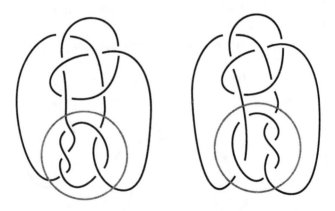

Figure 10.8

Show that mutant knots have the same Conway polynomial. (Hint: Show that a sequence of crossing changes and smoothings can be carried out within the circle so that the resulting links and knots have the property that each is unchanged by mutation.)   Use this to prove that the Alexander polynomial of the $(p_1, p_2, \ldots, p_n)$-pretzel knot is independent of the order of the p_i.

2 New Polynomial Invariants

In 1985, Jones described a new polynomial invariant of knots and links which was able to distinguish knots with the same Alexander polynomial. Jones's work used braid descriptions of knots. It was soon seen, however, that this *Jones polynomial* could be computed, and defined, via a recursion formula similar to that for the Conway polynomial. Almost immediately, it was recognized that the two recursion formulas are special cases of a recursion formula defining a 2-variable polynomial of knots and links. This new polynomial, named the HOMFLY *polynomial* after the initials of some of its discoverers (Hoste, Ocneanu, Millett, Freyd, Lickorish, and Yetter) contains information that is missed by both the Jones and Conway polynomials.

The recursion relation for the HOMFLY polynomial, $P_L(\ell, m)$, is given by

$$\ell P_{L_+}(\ell, m) + \ell^{-1} P_{L_-}(\ell, m) = -m P_{L_s}(\ell, m).$$

As with the Conway polynomial, this formula, along with the condition that for the unknot U, $P_U(\ell, m) = 1$, yields a well-defined polynomial link invariant.

EXAMPLES
The calculation for the unlink based on Figure 10.2 now

shows that the unlink of two components has polynomial $-m^{-1}(\ell + \ell^{-1})$. Letting

$$\mu = -m^{-1}(\ell + \ell^{-1}),$$

one can quickly show that the unlink of n components has polynomial μ^{n-1}. In general, the calculation of P proceeds in the exact same way as for the Conway polynomial. A few more examples are left as exercises for the reader:

$$P_{3_1}(\ell, m) = (-2\ell^2 - \ell^4) + \ell^2 m^2,$$
$$P_{4_1}(\ell, m) = (-\ell^{-2} - 1 - \ell^2) + m^2,$$
$$P_{6_2}(\ell, m) = (2 + 2\ell^2 + \ell^4) + (-1 - 3\ell^2 - \ell^4)m^2 + \ell^2 m^4.$$

As was demonstrated in Exercise 1.6, the orientation of the components of a link affect the value of the polynomial. This adds to the care required to do the calculations correctly.

EXERCISES

2.1. Carry out the calculation of the HOMFLY polynomial for the trefoil knot and its mirror image. Exercise 2.6 discusses the relationship between the polynomial of a knot and its mirror image.

2.2. Compute the HOMFLY polynomial of the knots 4_1 and 6_2.

2.3. Show that $\nabla(z) = P(i, iz)$, where $i^2 = -1$.

2.4. Use the sequence of crossing changes to compute the HOMFLY polynomial of the knot in Figure 10.5 above (see p. 215).

2.5. Show that the 11-crossing knot discussed in Exercise 1.5 has nontrivial HOMFLY polynomial.

2.6. Show that the HOMFLY polynomial of a knot and its mirror image are related by replacing ℓ by ℓ^{-1}.

3 Kauffman's Bracket Polynomial

The theory of polynomial invariants of knots continues to develop. Among the most significant advances is a new approach introduced by Kauffman. The Kauffman bracket polynomial is easy to define, and the proof that it is a knot invariant follows readily from the Reidemeister moves.

In an *unoriented* link diagram, D, crossings can be rotated to appear as in Figure 10.9a. Each crossing can then be smoothed in one of two ways, one of which is called a smoothing of type A and the other of type B, as

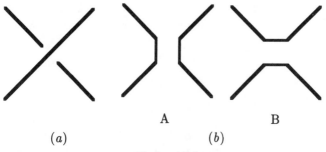

A   B

(a) (b)

Figure 10.9

indicated in Figure 10.9b. Kauffman defines a *state*, S, to be a choice of smoothings for each of the crossings in the diagram. For each state (if there are n crossings there are 2^n states), set $\langle D|S \rangle = t^{a-b}$, where a is the number of smoothings of type A and b is the number of type B smoothings. Next define

$$\langle D \rangle = \sum \langle D|S \rangle (-t^{-2} - t^2)^{|S|-1},$$

where the sum is taken over all states, and $|S|$ is the number of circles that result after all the smoothings of the given state are performed to the diagram.

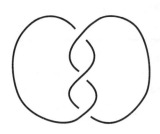

Figure 10.10

As an example, consider the diagram D for the trefoil in Figure 10.10. There is a total of eight states. One of the states has no type B changes. Three states have exactly 1 type B change. Another 3 states have exactly 2 type B changes. The last state has 3 type B changes. The resulting diagrams have $|S|$ equaling 2, 1, 2, and 3 respectively.

The resulting polynomial is shown below:

$$\begin{aligned}
\langle D \rangle &= t^3(-t^{-2} - t^2) + 3t \\
&\quad + 3t^{-1}(-t^{-2} - t^2) + t^{-3}(-t^{-2} - t^2)^2 \\
&= t^{-5} + t^{-3} + t^{-7}.
\end{aligned}$$

In general, the polynomial $\langle D \rangle$ can be shown to be invariant under Reidemeister moves 2 and 3, but it definitely changes when the first Reidemeister moves is performed. (As an easy but valuable exercise, what happens when a Reidemeister move 1 is performed to a trivial diagram for the unknot? There are two cases to consider, one where the added crossing is right-handed and the other when it is left-handed.)

To arrive at a polynomial that is invariant under all three Reidemeister moves, a correction term can be included as follows. Orient each component of the original diagram D, and call the resulting link K. Let w denote the number of right-handed crossings minus the number of left-handed crossings in the resulting oriented diagram. The Kauffman polynomial, $F[K]$, is defined to be $(-t)^{-3w}\langle D \rangle$. (The reader can easily verify that Reidemeister move 1 changes $\langle D \rangle$ by $(-t)^{\pm 3}$.) In the case of the trefoil, illustrated in Figure 10.10, $w = 3$, and the resulting polynomial is $t^{-4} + t^{-12} - t^{-16}$.

As in the example of the trefoil, the exponents of the resulting polynomial are always divisible by 4. Hence, $F[K](t^{-1/4})$ is a polynomial (in t and t^{-1}) and Kauffman proved that $F[K](t^{-1/4})$ is in fact the Jones polynomial. This new approach to the Jones polynomial is strikingly simple. More important, it can be used to define other invariants which have proved especially useful, and also leads to new insights into the Jones polynomial.

The Kauffman polynomial has proved especially useful in the study of combinatorial properties of knots. For instance, Kauffman, and independently Murasugi and Thistlethwaite, proved that if a knot has an alternating diagram, then all of its minimal crossing diagrams are alternating. A related result is the additivity of crossing number for alternating knots. As with the Alexander polynomial, the new knot polynomials reflect symmetries of knots and links, and these connections yield a variety of corollaries. Several excellent recent surveys concerning these new methods and results in knot theory are listed in the references.

Many questions remain open. There are knots which have trivial Alexander polynomial, but as of yet no knot has been found that cannot be distinguished from the un-knot using more general knot polynomials. More important, finding noncombinatorial interpretations of these new invariants is now a major area of research.

Exercises

3.1. Compute $F[K]$ where K is the (a) figure 8 knot, (b) (2,2)-torus link, (c) (2,−2)-torus link.

APPENDIX A: KNOT TABLE

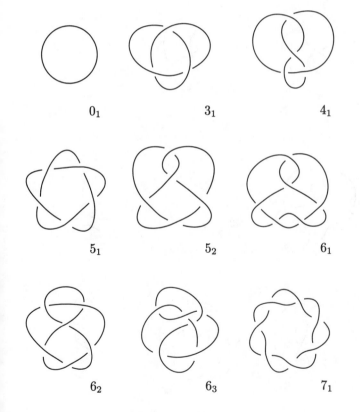

0_1 3_1 4_1

5_1 5_2 6_1

6_2 6_3 7_1

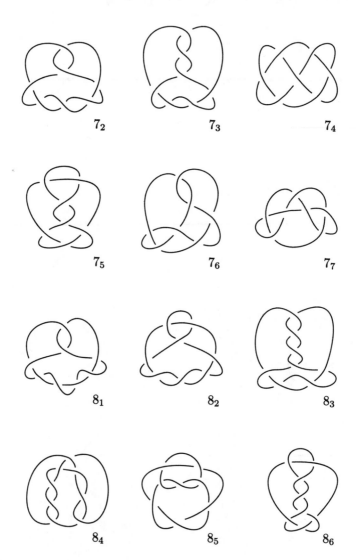

7_2 7_3 7_4

7_5 7_6 7_7

8_1 8_2 8_3

8_4 8_5 8_6

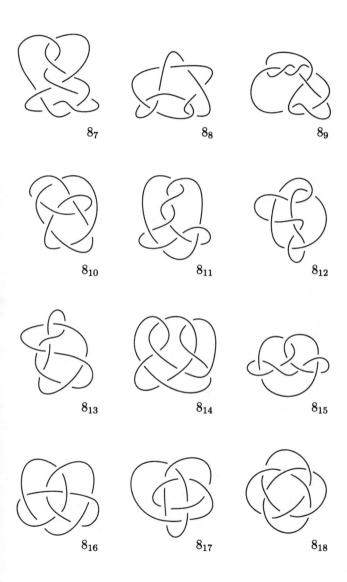

8_7

8_8

8_9

8_{10}

8_{11}

8_{12}

8_{13}

8_{14}

8_{15}

8_{16}

8_{17}

8_{18}

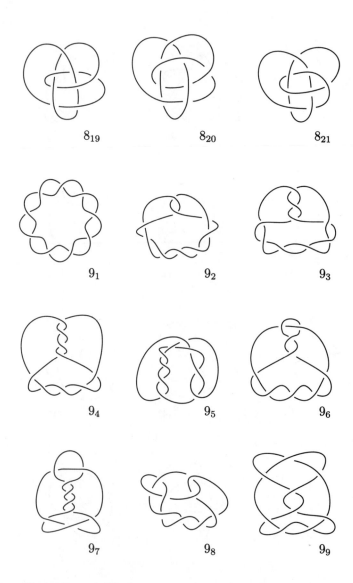

8_{19} 8_{20} 8_{21}

9_1 9_2 9_3

9_4 9_5 9_6

9_7 9_8 9_9

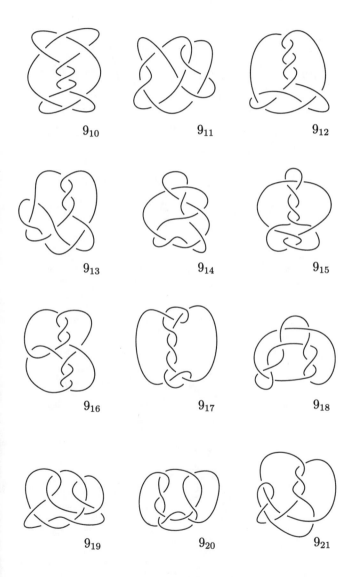

9_{10} 9_{11} 9_{12}

9_{13} 9_{14} 9_{15}

9_{16} 9_{17} 9_{18}

9_{19} 9_{20} 9_{21}

9_{22} 9_{23} 9_{24}

9_{25} 9_{26} 9_{27}

9_{28} 9_{29} 9_{30}

9_{31} 9_{32} 9_{33}

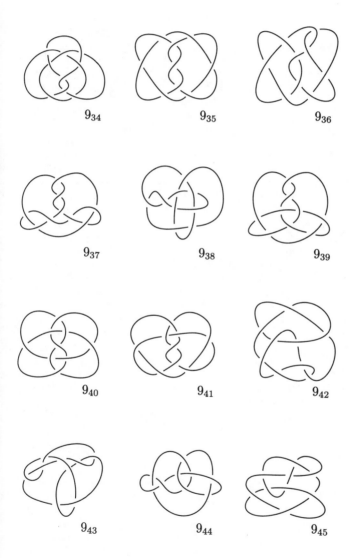

9_{34} 9_{35} 9_{36}

9_{37} 9_{38} 9_{39}

9_{40} 9_{41} 9_{42}

9_{43} 9_{44} 9_{45}

9_{46}

9_{47}

9_{48}

9_{49}

APPENDIX 2
ALEXANDER POLYNOMIALS

3_1	$t^2 - t + 1$
4_1	$t^2 - 3t + 1$
5_1	$t^4 - t^3 + t^2 - t + 1$
5_2	$2t^2 - 3t + 2$
6_1	$2t^2 - 5t + 2$
6_2	$t^4 - 3t^3 + 3t^2 - 3t + 1$
6_3	$t^4 - 3t^3 + 5t^2 - 3t + 1$
7_1	$t^6 - t^5 + t^4 - t^3 + t^2 - t + 1$
7_2	$3t^2 - 5t + 3$
7_3	$2t^4 - 3t^3 + 3t^2 - 3t + 2$
7_4	$4t^2 - 7t + 4$
7_5	$2t^4 - 4t^3 + 5t^2 - 4t + 2$
7_6	$t^4 - 5t^3 + 7t^2 - 5t + 1$
7_7	$t^4 - 5t^3 + 9t^2 - 5t + 1$
8_1	$3t^2 - 7t + 3$
8_2	$t^6 - 3t^5 + 3t^4 - 3t^3 + 3t^2 - 3t + 1$
8_3	$4t^2 - 9t + 4$
8_4	$2t^4 - 5t^3 + 5t^2 - 5t + 2$
8_5	$t^6 - 3t^5 + 4t^4 - 5t^3 + 4t^2 - 3t + 1$

8_6	$2t^4 - 6t^3 + 7t^2 - 6t + 2$
8_7	$t^6 - 3t^5 + 5t^4 - 5t^3 + 5t^2 - 3t + 1$
8_8	$2t^4 - 6t^3 + 9t^2 - 6t + 2$
8_9	$t^6 - 3t^5 + 5t^4 - 7t^3 + 5t^2 - 3t + 1$
8_{10}	$t^6 - 3t^5 + 6t^4 - 7t^3 + 6t^2 - 3t + 1$
8_{11}	$2t^4 - 7t^3 + 9t^2 - 7t + 2$
8_{12}	$t^4 - 7t^3 + 13t^2 - 7t + 1$
8_{13}	$2t^4 - 7t^3 + 11t^2 - 7t + 2$
8_{14}	$2t^4 - 8t^3 + 11t^2 - 8t + 2$
8_{15}	$3t^4 - 8t^3 + 11t^2 - 8t + 3$
8_{16}	$t^6 - 4t^5 + 8t^4 - 9t^3 + 8t^2 - 4t + 1$
8_{17}	$t^6 - 4t^5 + 8t^4 - 11t^3 + 8t^2 - 4t + 1$
8_{18}	$t^6 - 5t^5 + 10t^4 - 13t^3 + 10t^2 - 5t + 1$
8_{19}	$t^6 - t^5 + t^3 - t + 1$
8_{20}	$t^4 - 2t^3 + 3t^2 - 2t + 1$
8_{21}	$t^4 - 4t^3 + 5t^2 - 4t + 1$
9_1	$t^8 - t^7 + t^6 - t^5 + t^4 - t^3 + t^2 - t + 1$
9_2	$4t^2 - 7t + 4$
9_3	$2t^6 - 3t^5 + 3t^4 - 3t^3 + 3t^2 - 3t + 2$
9_4	$3t^4 - 5t^3 + 5t^2 - 5t + 3$
9_5	$6t^2 - 11t + 6$
9_6	$2t^6 - 4t^5 + 5t^4 - 5t^3 + 5t^2 - 4t + 2$
9_7	$3t^4 - 7t^3 + 9t^2 - 7t + 3$
9_8	$2t^4 - 8t^3 + 11t^2 - 8t + 2$
9_9	$2t^6 - 4t^5 + 6t^4 - 7t^3 + 6t^2 - 4t + 2$
9_{10}	$4t^4 - 8t^3 + 9t^2 - 8t + 4$

9_{11} $\quad t^6 - 5t^5 + 7t^4 - 7t^3 + 7t^2 - 5t + 1$

9_{12} $\quad 2t^4 - 9t^3 + 13t^2 - 9t + 2$

9_{13} $\quad 4t^4 - 9t^3 + 11t^2 - 9t + 4$

9_{14} $\quad 2t^4 - 9t^3 + 15t^2 - 9t + 2$

9_{15} $\quad 2t^4 - 10t^3 + 15t^2 - 10t + 2$

9_{16} $\quad 2t^6 - 5t^5 + 8t^4 - 9t^3 + 8t^2 - 5t + 2$

9_{17} $\quad t^6 - 5t^5 + 9t^4 - 9t^3 + 9t^2 - 5t + 1$

9_{18} $\quad 4t^4 - 10t^3 + 13t^2 - 10t + 4$

9_{19} $\quad 2t^4 - 10t^3 + 17t^2 - 10t + 2$

9_{20} $\quad t^6 - 5t^5 + 9t^4 - 11t^3 + 9t^2 - 5t + 1$

9_{21} $\quad 2t^4 - 11t^3 + 17t^2 - 11t + 2$

9_{22} $\quad t^6 - 5t^5 + 10t^4 - 11t^3 + 10t^2 - 5t + 1$

9_{23} $\quad 4t^4 - 11t^3 + 15t^2 - 11t + 4$

9_{24} $\quad t^6 - 5t^5 + 10t^4 - 13t^3 + 10t^2 - 5t + 1$

9_{25} $\quad 3t^4 - 12t^3 + 17t^2 - 12t + 3$

9_{26} $\quad t^6 - 5t^5 + 11t^4 - 13t^3 + 11t^2 - 5t + 1$

9_{27} $\quad t^6 - 5t^5 + 11t^4 - 15t^3 + 11t^2 - 5t + 1$

9_{28} $\quad t^6 - 5t^5 + 12t^4 - 15t^3 + 12t^2 - 5t + 1$

9_{29} $\quad t^6 - 5t^5 + 12t^4 - 15t^3 + 12t^2 - 5t + 1$

9_{30} $\quad t^6 - 5t^5 + 12t^4 - 17t^3 + 12t^2 - 5t + 1$

9_{31} $\quad t^6 - 5t^5 + 13t^4 - 17t^3 + 13t^2 - 5t + 1$

9_{32} $\quad t^6 - 6t^5 + 14t^4 - 17t^3 + 14t^2 - 6t + 1$

9_{33} $\quad t^6 - 6t^5 + 14t^4 - 19t^3 + 14t^2 - 6t + 1$

9_{34} $\quad t^6 - 6t^5 + 16t^4 - 23t^3 + 16t^2 - 6t + 1$

9_{35} $\quad 7t^2 - 13t + 7$

9_{36} $\quad t^6 - 5t^5 + 8t^4 - 9t^3 + 8t^2 - 5t + 1$

9_{37}	$2t^4 - 11t^3 + 19t^2 - 11t + 2$
9_{38}	$5t^4 - 14t^3 + 19t^2 - 14t + 5$
9_{39}	$3t^4 - 14t^3 + 21t^2 - 14t + 3$
9_{40}	$t^6 - 7t^5 + 18t^4 - 23t^3 + 18t^2 - 7t + 1$
9_{41}	$3t^4 - 12t^3 + 19t^2 - 12t + 3$
9_{42}	$t^4 - 2t^3 + t^2 - 2t + 1$
9_{43}	$t^6 - 3t^5 + 2t^4 - t^3 + 2t^2 - 3t + 1$
9_{44}	$t^4 - 4t^3 + 7t^2 - 4t + 1$
9_{45}	$t^4 - 6t^3 + 9t^2 - 6t + 1$
9_{46}	$2t^2 - 5t + 2$
9_{47}	$t^6 - 4t^5 + 6t^4 - 5t^3 + 6t^2 - 4t + 1$
9_{48}	$t^4 - 7t^3 + 11t^2 - 7t + 1$
9_{49}	$3t^4 - 6t^3 + 7t^2 - 6t + 3$

REFERENCES

Books and Survey Articles Six texts on knot theory are listed here. The book by Reidemeister, though dated, offers an accessible account of many of the details not included here. Crowell and Fox focus on algebraic techniques, and include a careful presentation of the fundamental group. Basic algebraic topology is a prerequisite of Rolfsen's book, and Burde and Zieschang offer the most advanced treatment. Kauffman begins with an elementary presentation; the latter chapters call on a background in algebraic topology. Finally, Moran's book begins with a development of the basic tools of the subject, such as the fundamental group, and then focuses on the special topic of braids and the relation of braids to knots and links. Birman's book offers a more advanced treatment of braids.

Four survey articles are listed. The papers by Kauffman and by Lickorish and Millet are excellent introductions to the new combinatorial methods summarized in Chapter 10. Fox's "Quick Trip" includes a discussion of higher dimensional knots and also summarizes many of the methods and results of classical knot theory. The article by Gordon is an excellent survey from an advanced viewpoint.

References

Books

J.S. Birman, *Braids, Links, and Mapping Class Groups*, Annals of Mathematics Studies Vol. 82, Princeton University Press, Princeton, 1974.

G. Burde and H. Zieschang, *Knots*, de Gruyter Studies in Mathematics 5, Walter de Gruyter, Berlin-New York, 1985.

R. H. Crowell and R. H. Fox, *Introduction to Knot Theory*, Graduate Texts in Mathematics Vol. 57, Springer-Verlag, New York-Heidelberg-Berlin, 1977.

L. H. Kauffman, *On Knots*, Annals of Mathematics Studies 115, Princeton University Press, Princeton, 1987.

S. Moran, *The Mathematical Theory of Knots and Braids, An Introduction*, North-Holland Mathematical Studies 82, North-Holland, Amsterdam-New York-Oxford, 1983.

K. Reidemeister, *Knotentheorie*, Ergebnisse der Mathematic, Vol. 1, Springer-Verlag, Berlin, 1932; L. F. Boron, C. O. Christenson, and B. A. Smith, (English translation), BCS Associates, Moscow, Idaho, 1983.

D. Rolfsen, *Knots and Links*, Mathematics Lecture Series 7, Publish or Perish, Inc., Berkeley, 1976.

Survey Articles

R. H. Fox, *A quick trip through knot theory*, Topology of 3-Manifolds (M. K. Fort, Jr., ed.), Prentice-Hall, Englewood Cliffs, N. J., 1962.

C. McA. Gordon, *Aspects of classical knot theory*, Knot Theory, Lecture Notes in Math, Vol. 685, Springer-Verlag, New York, 1978.

L. H. Kauffman, *New invariants in the theory of knots*, Am. Math. Monthly **95** (1988), 195-242.

W.B.R. LICKORISH AND K. MILLETT, *The new polynomial invariants of knots and links*, Mathematics Magazine **61** (1988), 3-23.

RESEARCH ARTICLES AND OTHER REFERENCES

Original sources for results described in the book are listed below.

E. A. ABBOTT, *Flatland*, Dover Publications, New York, 1952.

J. W. ALEXANDER, *Topological invariants of knots and links*, Trans. Amer. Math. Soc. **30** (1928), 275-306.

J. W. ALEXANDER AND G. B. BRIGGS, *On types of knotted curves*, Ann. of Math. **(2) 28** (1927), 562-586.

E. ARTIN, *Theorie der Zöpfe*, Abh. Math. Sem. Univ. Hamburg **4** (1925), 47-72.

———, *Theory of braids*, Ann. of Math. **(2)48** (1947), 101-126.

S. A. BLEILER, *A note on unknotting number*, Math. Proc. Camb. Phil. Soc. **96** (1984), 469-471.

J. H. CONWAY, *An enumeration of knots and links, and some of their algebraic properties*, Computational Problems in Abstract Algebra, Proc. Conf. Oxford 1967 (J. Leech, ed.), Pergamon Press, New York, 1970.

M. DEHN, *Die beiden Kleeblattschlingen*, Math. Ann. **75** (1914), 402-413.

A. EDMONDS, *Least area Seifert surfaces and periodic knots*, Topology and its Apps. **18** (1984), 109-113.

P. FREYD, D. YETTER, J. HOSTE, W. B. R. LICKORISH, K. MILLETT, AND A. OCNEANU, *A new polynomial invariant of knots and links*, Bull. Amer. Math. Soc. **12** (1985), 239-246.

C. McA. GORDON AND J. LUECKE, *Knots are determined by their complements*, Journal of the Amer. Math. Soc. **2** (1989), 371-415.

V. F. R. JONES, *103-112*, Bull. Amer. Math. Soc. **12** (1985).

L. H. KAUFFMAN, *State models and the Jones polynomial*, Topology **26** (1987), 395-407.

M. KERVAIRE, *Knot cobordism in codimension 2*, Manifolds Amsterdam 1970, Lecture Notes in Math., Springer-Verlag, Berlin-Heidelberg-New York, 1971.

J. LEVINE, *Knot cobordism groups in codimension two*, Comm. Math. Helv. **45** (1969), 229-244.

W. H. MEEKS, AND S. T. YAU, *Topology of three dimensional manifolds and the embedding problem in minimal surface theory*, Ann. of Math. **112** (1980), 441-484.

K. MURASUGI, *On the genus of the alternating knot II*, J. Math. Soc. Japan **10** (1958), 235-248.

K. MURASUGI, *On periodic knots*, Comment. Math. Helv. **46** (1971), 162-174.

K. MURASUGI, *Jones polynomials and classical conjectures in knot theory I and II*, Topology **26** (1987), 187-194.

C. D. PAPAKYRIAKOPOULOS, *On Dehn's lemma and the asphericity of knots*, Ann. of Math. **66** (1957), 1-26.

K. PERKO, *On the classification of knots*, Proc. Amer. Math. Soc. **45** (1974), 262-266.

M. SCHARLEMANN, *Unknotting number one knots are prime*, Inv. Math. **82** (1985), 37-56.

O. SCHREIER, *Über die Gruppen $A^a B^b = 1$*, Abh. Math. Sem. Univ. Hamburg **3** (1924), 167-169.

H. SCHUBERT, *Die eindeutige Zerlegbarkeit eines Knotens in Primknoten*, S. Ber. Heidelberg, Akad. Wiss. **3. Abh.** (1949), 57-104.

———, *Über eine numerische Knoteninvariante*, Math. Z. **61** (1954), 245-288.

H. SEIFERT, *Über das Geschlect von Knoten*, Math. Ann. **110** (1934), 571-592.

P. G. TAIT, *On knots*, Scientific Papers, Cambridge University Press, London, 1898.

M. THISTLETHWAITE, *A spanning tree expansion of the Jones polynomial*, Topology **26** (1987), 297-309.

———, *Knot tabulations and related topics*, Aspects of Topology: In Memory of Hugh Dowker, 1912-1982 (I. M. James and E. H. Kronheimer, eds.), London Math. Soc. Lecture Notes Series 93, Camb. Univ. Press, Cambridge, 1985.

H. TROTTER, *Non-invertible knots exist*, Topology **2** (1964), 341-358.

F. WALDHAUSEN, *On irreducible 3-manifolds that are sufficiently large*, Ann. of Math. **87** (1968), 56-88.

W. WHITTEN, *Inverting double knots*, Pacific J. of Math. **97** (1981), 209-216.

E. C. ZEEMAN, *Twisting spun knots*, Trans. Amer. Math. Soc. **115** (1965), 471-495.

INDEX